CAREERS
WORKING WITH ANIMALS

Allan Shepherd

eighth edition

KOGAN PAGE

First published in 1981, author Helen Young
New editions 1983, 1986
New editions 1988, 1990 by Helen Young and Vivien Donald
Sixth edition 1992 by Vivien Donald
Seventh edition 1996 by Vivien Donald and Allan Shepherd
Eighth edition 1997 by Allan Shepherd

Apart from any fair dealing for the purposes of research or private study, or criticism or review, as permitted under the Copyright, Designs and Patents Act, 1988, this publication may only be reproduced, stored or transmitted, in any form or by any means, with the prior permission in writing of the publishers, or in the case of reprographic reproduction in accordance with the terms of licences issued by the Copyright Licensing Agency. Enquiries concerning reproduction outside those terms should be sent to the publishers at the undermentioned address:

Kogan Page Limited
120 Pentonville Road
London NI 9JN

© Kogan Page Ltd 1981, 1983, 1986, 1988, 1990, 1992, 1996, 1997

British Library Cataloguing in Publication Data

A CIP record for this book is available from the British Library.

ISBN 0 7494 2386 2

Typeset by Northern Phototypesetting Co Ltd, Bolton
Printed and bound in Great Britain by
Clays Ltd, St Ives plc

Contents

1. **Introduction**   1
Is this the job for you?

2. **Vets**   5
Introduction; Other prospects; Salaries; Qualities

3. **Veterinary nurses**   12
Qualifications; Training; Prospects

4. **Animal welfare and conservation charities**   17
Royal Society for the Prevention of Cruelty to Animals; People's Dispensary for Sick Animals; The Blue Cross; Animal conservation; RSPB; Animal sanctuaries

5. **Working with horses**   25
Stable lads; Stud work; Show jumping; Riding schools; British Horse Society; National Vocational Qualifications/Scottish Vocational Qualifications; The Association of British Riding Schools; Farriers; The armed services

6. **Working with dogs**   36
Kennel work; Greyhound racing; Dog training; Police dogs; Guide dogs; Hearing Dogs for the Deaf; Guard dogs

7. **Pets**   46
Pet shops; Animal grooming; Animal photography

Contents

8. Zoos 51
Introduction; Zoo keeping; Qualifications and training; What the job is really like; Visitors; Further qualifications; Salaries; Making a start

9. Agriculture 58
Training; Fish farming; Organic farming; City farms

10. Getting in and getting on 66
Top tips for getting into animal work

11. The future of animal work 68

12. Qualifications available 71

13. Where to study 75
Section one: degree courses in veterinary science; Section two: college courses; Section three: correspondence courses

14. Useful addresses 83

15. Further reading 88

Index *90*

1 Introduction

Is this the Job for You?

- ❒ Is it important to you to spend most of your working life with animals?
- ❒ Have you spent any length of time caring for an animal?
- ❒ Are you prepared to spend time learning skills and training for a qualification?
- ❒ If you have to, are you prepared to accept low wages as a price worth paying for job satisfaction?
- ❒ Are you hard working?
- ❒ Are you physically fit?
- ❒ Do you want to work outdoors?
- ❒ What sort of relationship would you like to form with the animals you work with?
- ❒ Are you prepared to put the animals in your care first, before any of your own considerations?
- ❒ Can you cope with the demands animals make without losing your temper or inflicting cruelty on the animals you work with?

Our relationship with animals is a complex one. We easily distinguish between those we keep as pets, those we use as working animals, those we race, those we eat, those which help us to protect or enhance our lives and those which we protect in their natural habitat. Then there are those animals that others have neglected, bullied and abused, and that need a special kind of care.

This book reflects the eclectic nature of this broad employment area and provides information on a wide range of careers, from agriculture to zoos. For each job mentioned there is a description of the type of training you need, the sort of work you will be doing, the opportunity for advancement and the level of wage you might expect when starting out.

Also listed are the various organisations that help individuals to fulfil their aspirations by offering in-depth advice on careers and courses. If this book is the first step towards finding a career working with animals then these organisations will help you to move on to the next stage, be it finding work experience, choosing a course, starting a modern apprenticeship or in some cases getting your first paid job.

Each chapter contains a number of case studies. This is your opportunity to find out what the people who already work with animals have to say about their jobs, their lives and their aspirations. These case studies can help you decide whether or not a job seems right for you. Take this example; 'Looking after a herd means getting up about 4am and you can't just go back to bed if it's raining or you had a heavy night the night before.' You'll know straight away whether this is you or not.

With such a wide range of careers to choose from, how will you decide which one is right for you? Sometimes it will be obvious, like the example above. Sometimes the decision will be made for you; a jockey, for example, must not weigh more than 8 stones/51 kg. Other times it will be less clear cut. A veterinary nurse could work for an animal welfare charity or a private veterinary surgeon and still perform the same basic duties.

And while some careers offer new recruits on-the-job training and a chance to gain NVQs, others demand academic ability, proved through long periods at university. Vets, for example, need three A-levels at grades AAB just to be in the running for a place at veterinary college.

Most people start off by having an interest in a particular species and go on to turn their interest into a career. If you already know which animal you like best, choosing a career is made considerably easier.

But even within a species there are a number of related careers. Horse lovers could choose between racing, breeding, jumping, instruction, welfare, therapy or veterinary surgery. Dog lovers could choose between training, breeding, kennel work and racing and then if they choose training between pets, guide dogs, police dogs, army dogs and so on.

Some people choose an area of work instead, and begin to specialise later. Zoo keepers may already have a specialist knowledge but many will start off as volunteers and develop a keen interest in one particular species. The same is true of agriculture. While some farmers will know before they start training that they want to work with sheep or pigs or cows, others may choose while at agricultural college.

Nobody really knows why people develop an affinity for a particular species, but most people stick to one choice throughout their careers. This is partly because training is often biased towards one choice so that employees and employers get the best deal from any investment of time and money they make. One of the exceptions to this rule is veterinary surgery and nursing where trainees have to learn about the specific needs of a number of animals.

While many people crave a job working with animals, few who have experienced this rewarding work would swap it for any other. Although wages are generally lower and the work more physically demanding than in many other jobs, most people who work with animals would readily echo the words of this farm worker, 'You don't get drained doing this work. Exhausted, yes, but a healthy exhaustion'.

Anyone who thinks they want to work with animals should try it out first. Although surprisingly few are put off by the initial impression, if you aren't going to like certain aspects of the job it's best to find out sooner rather than later.

All animal work calls for some sort of vocation but this takes many forms. The desire to care for animals in trouble, which leads to nursing and charity work, is quite different from the instinct to control which is needed by trainers of show jumpers.

Vets may well be animal lovers, but their attitude to them is much like that of a good GP towards patients, emotionally detached. That is almost the opposite of the approach needed by a rider whose involvement with a horse is total.

Choosing which of the relationships suits you best is the key to finding the right job.

2 Vets

Introduction

While many youngsters would list becoming a vet as their first ambition, the high academic requirements, shortage of places and length of courses ensure that only a small proportion of these become qualified veterinary surgeons. The six universities which teach veterinary science – Bristol, Cambridge, Edinburgh, Glasgow, Liverpool and London – all demand A-level or Higher grades AAB in subjects such as chemistry, physics, biology and zoology. Even then, an applicant with these grades may not get a place as only 400 students are accepted to train each year. Any would-be student should write to the universities, which are listed in Chapter 10, to find out what qualities and qualifications they are looking for. The Royal College of Veterinary Surgeons produces a leaflet *The Veterinary Profession* (free), a detailed booklet *Careers for Veterinary Surgeons* (£4.00) and a video *Do You Want to be a Vet?* (£16.00). Another useful book is *The Veterinary Science Degree Course Guide* from CRAC Publications.

The veterinary schools' tradition of picking the brightest students has its critics, who feel it is bound to produce too many academics and not enough general practitioners. There is no doubt that it does exclude a few who might have made excellent working vets. However, it is unlikely to change, and it is defended by most qualified vets including one whose own 20-year-old son, who had three A-levels, was turned down. He said: 'This is a branch of medicine and standards have to be high. Vet schools have always produced a few academics, and who else would do the research and the teaching? But people who go to vet school wanting to be vets still want to be vets when they qualify.'

Case Study

One vet who excelled as a student is now in practice in Fife.

'My father was a vet and I grew up with the life. I always knew I would go into his practice one day. Certainly, nothing I experienced at vet school changed my mind, though it happened that I did quite well there. It is certainly the case that a professor will try to make an academic of a good student. My prof worked on me to stay on for another two years to do a BSc.

Well, it's flattering, of course, and it can also make you wonder. I agreed at one point and did stay on for almost a year, but my heart wasn't in it.

I had always wanted to be a working vet, doing what I'm doing now, so I left. I'm perfectly satisfied with my life. It is different perhaps in that I have become an expert in one particular species, and that has taken me round the world a bit, as adviser or giving technical evidence in disputes and court cases. But most of my life is spent here, about 50 per cent of my time with horses and cows belonging to local families I've known all my life. The rest of the work is with dogs and cats. You'll find few practices now, even in farm country, which don't handle small animals.

I suppose the life is pretty much what you'd expect if you read James Herriot. I can't fault those books – in fact I wish I'd written them. The big difference, of course, is that they were set in the thirties, before reliable anaesthetics and, most important, before antibiotics. A vet's life in those respects is easier today. But have you noticed, when James Herriot went to a calving there was always a queue of people there, waiting to help? When I turn out there's a farmer with a bad back or something, and his wife who is just going to bed. But the routine *is* much the same, the same variety and triumphs and disasters.'

It is not, however, the life for all vets. There are over 14,000 in Britain and only a little more than 7,000 work in private practice.

Case Study

Eileen *is a vet in a Grimsby animal hospital. She has three nurses working for her whom she trained herself, and extremely well-equipped, modern working surroundings.*

'I have worked in private practice, but I didn't particularly like it. The work I'm doing now is entirely with small animals, and I certainly find it just as interesting and fulfilling. It's a question of being practical. Who wants to go out in some mucky farmyard?

It's much nicer to stay in a centrally heated building than to lie on your belly in the mud. No, I don't think this is anything to do with being female. There are a great many women vets, and I'm sure they can tackle any job the men can. A few, if they're on the small, dainty side, might need a hand – say with some cattle work. But if private practice appeals to them, they'll do it just as well as a male vet. It's a matter of personal preference.'

Thanks to research, there are constant developments and improvements in veterinary medicine. A busy practitioner *has* to keep up to date, but has difficulty in finding time for any of the innumerable refresher courses available. This is one reason why most vets now work in group practices. Another is the anti-social hours of the job.

Case Study

One West Country vet began on his own more than 20 years ago, and now shares a practice with three others.

'You can still stick your plate up and start off, but it would just about kill off a vet nowadays. A vet has to supply a 24-hour service. There has to be someone always on call, and when he's out there has to be someone there to answer the telephone. No vet can count on a social life. You might invite your friends round for the evening, then arrive half-way through dripping of goodness knows what and smelling to high heaven.

Even without emergencies, the hours are anti-social. Even in group practices, there's surgery three evenings a week until 7 pm and two or three nights a week on call.

Different aspects of the work appeal more to some than others. In my practice routine surgery is done five mornings a week by one partner. The sort of thing he's doing is spaying and castrating cats and dogs, and operating on accident victims and sick animals with tumours. There's a lot of work on teeth and ears. Now for some, this is the most interesting side of the work. To me everything is exciting the first dozen times.

The first time you spay a cat – there's never been surgery like it! It's probably the first operation you've done entirely alone. No, you're not frightened. You're totally confident. It's the third or fourth time, when you know a bit more about it, that you're frightened.

Veterinary work is a lot about people, and for me that is becoming more and more important and interesting. I'm a bit of a missionary about this. It may well be that when you see Mrs Bloggins' cat with the abscess for the tenth time it begins to lose its fascination – for the vet, but not for Mrs Bloggins. She's worried about her cat and she needs reassurance. There are hundreds of Mrs Blogginses. I feel it's a perfectly good use of veterinary time to reassure the humans. They're at least as important as the animals.'

Other prospects

The State Veterinary Service (SVS)

The SVS advises the Ministry of Agriculture, Fisheries and Food on the maintenance of the maximum level of health in British farm animals. It comprises administrative centres in Surrey, Edinburgh and Cardiff and the Veterinary Field Service (VFS). The VFS has a wide range of responsibilities including notifiable disease control, control of the importation of animals, enforcement of farm animal welfare legislation and meat hygiene monitoring.

There are five regional offices with jurisdiction over smaller divisions. Each division has its own Animal Health Office, staffed by a team of veterinary officers, technical staff and support staff.

Veterinary officers primarily work in one of the five regional organisations and are attached to a divisional office depending on their inclinations, aptitudes and the existence of vacancies. Their duties will include the control and eradication of notifiable disease, disease monitoring and surveillance, special investigations and surveys, red meat and poultry hygiene and the pursuit of improved animal welfare. Candidates for these positions must be members of the Royal College of Veterinary Surgeons.

Animal health officers assist veterinary staff with their duties and carry out non-veterinary work such as brucellosis testing and residue collection. Most of the work is done on farms. The minimum entry requirement is four GCSE passes or equivalent including English and a scientific subject. As competition is keen any additional agricultural qualification would be an advantage.

All appointments to the State Veterinary Service are made by the Recruitment and Assessment Service.

Animal welfare

Many vets prefer hospital work and some are employed by animal welfare organisations, such as the PDSA (People's Dispensary for Sick Animals) which – as Britain's largest veterinary charity – employs over 200 veterinary surgeons. It is not necessary to have a special vocation for charitable work, although many vets clearly do. In terms of small animal practice experience the PDSA provides a wide range of opportunities. The standard of care provided matches that of a good quality private practice. For more information see Chapter 3.

Inspection

Vets are also employed in poultry processing plants, doing meat and hygiene inspection. They are in charge of welfare and general inspection, such as water and chilling temperatures. They make sure that animals are properly stunned and killed humanely.

'Under the Meat Hygiene Regulations large plants have to employ a vet. Some do this sort of work part-time, checking abattoirs and cold stores, import and export depots, certifying that standards are being met. One can take pride that a veterinarian certificate is valued as fair.'

No matter what sort of work attracts the qualified vet, each student must spend six months working in a practice, and most start between their third and fourth years, during holiday periods.

Case Study

Harry is a fourth-year veterinary student.

'I've just spent Christmas in a practice in the north and thoroughly enjoyed it. I lived with one of the partners' family and that side of it was really great, especially New Year's Eve. You could tell just by looking round the room who was on call. It was my second spell in practice and you don't exactly lead a sheltered life at vet college, so I knew what to expect, but what I didn't reckon on was the cold.

Farmers seem to be one extreme or the other. Some call in the vet because they are vaguely worried about an animal when it's much too soon to make a diagnosis. We had one like that and the senior partner turned to me, straight-faced, and said, 'What would *you* say, Harry?' Naturally I thought he had it all sussed out and I couldn't see a damn thing wrong with the brute. Others seem to wait until the animal is about to die, which is a bit depressing because by then it's too late to do much.

In this practice there was a farmer whose panic level was pretty low. If he rang to say a cow was about to calve, the vet would say, 'Better get a move on or it will arrive without us and there will be no fee.' Most cows get on with it, without help, but about 5 per cent do need assistance. I've never been in on any of the dramatic midnight calvings by storm lantern so much loved in fiction. But I did go on a few night calls.

There are four vets in that practice and they take night duty in turns, but a student is supposed to come on all enthusiastic and not miss a thing.

Actually, to be quite serious, I wouldn't have missed any of it. But there was an hour or so, freezing in a pig shed, wearing someone else's gumboots which were giving me hell because my own were still soaked from the last call, when I asked myself if maybe I should be looking for a cushier job like, say, oil-rig drilling in the North Sea.'

Salaries

A newly qualified veterinary surgeon entering general practice as an assistant will receive around £15,500–£21,500 including car and accommodation, and a continued professional development allowance, rising to about £25,000 according to experience. Small-animal practices are generally considered to be the most remunerative (£30,000 for a clinician with appropriate specialisations), especially for those who eventually become partners. A new veterinary graduate on a university's staff would receive around £12,900. MAFF scales start at £20,000 to £41,120 at Grade 5 (1997) and commercial company rates are usually similar to MAFF salaries, up to around £30,000.

Qualities

The demands made by universities are understandable. The amount of academic work covered by them is formidable but, just like doctors, vets need other qualities more difficult to measure. All through their years of study they are coming into contact with live animals. Their training gears them to cure and prevent disease. The realities of practice call for a heavy blend of caring and detachment. The responsibilities they take on are a high price to pay for the lifestyle most of them will say is ideal.

3 Veterinary nurses

No vet could survive for long in practice without assistants. In the words of a 19-year-old who works in a London practice: 'There's a girl who keeps the records, a girl who makes appointments, a receptionist, a pre-op and a post-op nurse, and a girl who makes the tea. Me.'

In fact there are two trainee nurses in that practice and they both perform all the duties she described. They also help with X-rays, anaesthetics and sterilising of instruments and do a very large amount of cleaning up. They will learn about theatre work, ordering drugs, laboratory diagnostic tests, and the care and treatment of animal patients kept in the ward. They are training for qualification as Veterinary Nurses (VNs).

The British Veterinary Nursing Association (BVNA) controls training standards and anyone considering a career as a veterinary nurse should contact them first. On receipt of an sae they will send out their information pack. If you want full details of the training scheme, a *Guide to the Scheme* is available from the BVNA for £4.80.

Qualifications

A trainee needs four A, B or C passes at GCSE/SCE, which must include English language and either a physical or biological science or mathematics, or appropriate passes in examinations of a comparable or higher standard. If you do not have the appropriate qualifications it is still possible to take a pre-veterinary nursing qualification, which is the equivalent of four GCSE/SCE passes. Training takes a minimum of two years. An applicant must be at least 17 (there is no upper limit). There are two examinations to be passed before entry to the VN Register.

Training

The first step is to find employment at an Approved Training Centre (ATC). Without this, or the promise in writing of such a job, no one can enrol with the Royal College of Veterinary Surgeons (RCVS) as a trainee. These centres are veterinary practices or animal centres, such as those run by the RSPCA, PDSA and the Blue Cross, which have been approved by the Royal College of Veterinary Surgeons. A list is available from the BVNA.

Posts for trainees are advertised weekly in the *Veterinary Record* (you may be able to borrow copies from your local vet's surgery or from the library) and in the *Veterinary Nursing Journal*. This is available from the BVNA.

You may also be able to find a place through the Employment Register set up by the BVNA. This is a computer-based register of those looking for work in veterinary surgeries, which is sent out to surgeries and animal hospitals each week. Write for a form to the BVNA at the address above, enclosing an sae.

Training extends over a minimum of two years and combines on the job training with college instruction and learning through books. A trainee must have completed two years' training and passed Part I and Part II examinations to qualify as a Veterinary Nurse. A list of colleges offering VN courses appears in Chapter 13. Enrolment in such courses, which usually start in September, does not constitute enrolment as a VN trainee, for which a separate application must be made to the RCVS, through the BVNA. Berkshire, Bicton (Devon), Kingston Maurward (Dorset), Kirkley Hall (Northumberland) and Rodbaston (Staffordshire) Colleges of Agriculture, University of Bristol School of Veterinary Science and Wood Green Animal Shelters (Cambridgeshire) all offer full-time residential courses. All other colleges listed offer part-time courses.

Working for a local veterinary surgery on a voluntary basis gives a chance to get an idea of the duties of a veterinary nurse, and the experience can help in getting a full-time place.

Case Study

Kim is 20 and works for a practice in Hertfordshire.

'I wanted to be a vet, but I wasn't sure I wanted to stay on for my A-levels, so I decided to be a nurse.

I did a bit of work with animals first, to find out if I was as keen as I thought, and I recommend that to anyone who is thinking of taking this up. I worked for eight months as a kennel maid, then applied for a place in an approved practice.

The course is actually a small animals course, but the practice I'm with is mixed, dealing with horses, cows, pigs and sheep as well as cats and dogs, so my life is a lot more varied than most. I go out to farms with the vet. I also do a bit of reception work, give out drugs, answer telephones – I book appointments, making decisions about what is an emergency.

We have a week on the ward, mainly post-operative care of patients, some with long-standing diseases. Sick animals can get a bit snappy but it doesn't bother me. If a dog is determined to have a go I'll bandage up its nose.

I assist in ops, taking care of sterilising instruments, administering anaesthesia – and cleaning up. There's a lot of that and it gets very messy. Vomit, diarrhoea, blood – not for the squeamish.

I am in my second year now and I earn about £50 a week [rates may be higher depending on the area you live in]. It's not marvellous, but I have free accommodation. In fact a lot of people leave because of the money. It means you have to be dependent on your parents and although I get on very well with mine I'd rather not have to ask them for money at my age. Still, I suppose if I had gone to veterinary college I'd be round their necks a lot longer. Because so many people want to do this job, vets don't have to pay more.

I go to college once a week. The course covers anatomy and physiology of dogs, cats, birds – bones, muscles, nerves, liver, kidneys. I draw lots of lovely labelled diagrams. Actually, I am enjoying the studying side. I even thought I might, after all, go on to try for vet college, but I decided I enjoy my job more.'

Prospects

Trainees expect to earn between £3,000 and £7,000 a year depending on which region of the country they work in. After qualifying, a VN can expect to earn around £5,000–£9,000 per year, although there is no fixed salary structure and pay varies considerably in different parts of the country. Basic pay in a private practice may be £75 a week – less than £4,000 per year. Accommodation goes with one in five jobs, plus bonuses and overtime. Senior nurses earn up to £12,000–£13,500 a year, plus overtime. Most stay in general practice work, although many go on to work in hospitals or universities where salaries tend to be higher.

Opinions about VNs vary in the profession. One Scottish vet said: 'Frankly, I don't think their career prospects are too hot. I wouldn't hire one. I prefer to take on a local girl and train her myself. I want my staff to do things my way and not the way they've been taught in another practice and college. And to be quite candid, there are so many people keen to work with vets that we can afford to be choosy.'

A hospital vet who employs three trained VNs disagrees. 'Admittedly I trained all mine myself and they do very well here. They have a good salary scale. I think anyone who wants a trained VN has to pay properly these days. I am not entirely sure that VNs always get such a thorough training in practices, but there seems to be a great demand for them. There are advertisements every week in veterinary papers. I think they would have little trouble finding good, interesting jobs.'

Not all animal nurses want to train as VNs. There are vet practices all over the country employing untrained staff who take on most of the duties which would be done by a VN trainee. Vet practices are businesses and they need competent lay staff, but few can afford to pay people simply to answer the telephone and act as receptionists.

Case Study

Rosemary is 22 and works as a receptionist for a vet in Essex.

'I did a secretarial course and had quite a good job with a cosmetics firm for a year, but I was unsettled. It's the old story. I had always wanted to work with animals, but I knew I'd never stick all those years in college, so I gave up the idea. Then this job came up. It meant quite a drop in pay – I was earning a good salary in my old job. I was taken on as a receptionist and that's what I am, officially. I do work as a receptionist. We all do here, taking it in turns.

There are four of us, two vet's assistants and two lay staff, but we all find ourselves helping out in odd ways. The duty vet might ask one to help hold an animal during surgery.

I don't do any op work, although I help out with the cleaning up and sterilisation. The first time I asked to watch an operation I was a bit afraid I might blow it. It was a spaying – a lot of them are. There was quite a lot of blood and mess, but the strange thing is, I was too interested to feel sick or anything.

The thing that upsets me is when an animal is put down. If an animal is ill or dying, then I can see it's the only merciful thing to do. But people bring healthy dogs and cats here, just to get them out of the way. The vets I work for have a policy of refusing to put down a healthy animal, but you'd be surprised how many people ask. I suppose it's better than chucking them out to fend for themselves, but there's a look about those animals, and it's not just my imagination, that really breaks me up.'

4 Animal welfare and conservation charities

Animals throughout the world are cared for and protected by charities. Wild, domestic and farm animals which are deserted, neglected, ill-treated or tortured often die in misery. The lucky ones make it to an animal hospital or sanctuary. Wild animals face multiple threats, from hunting and trade to pollution and the destruction of their habitats.

While the need for animal care is great, the funds available to charities are not. With limited commercial opportunities most animal welfare charities survive by employing a mixture of paid staff and volunteers. For volunteers this is not necessarily a bad thing. Volunteering provides a valuable opportunity to get work experience and meet potential employers.

For some, animal welfare work is the only option. Many of the organisations listed in the *Animal Rescue Directory 1996/7* (available from Hand to Paw, price £2.95) were started by single-minded individuals dedicated to the protection of a particular species or type of animal. It is their vocation because the rewards of their work are the animals that go on to lead happier lives.

For many animals such help comes too late. Anyone thinking of taking up a job in animal welfare should be prepared for the sad days.

Royal Society for the Prevention of Cruelty to Animals

The RSPCA is the biggest animal charity. It operates 35 clinics, nine welfare centres, four hospitals and 50 animal homes. As well as 305 inspectors, it employs kennel assistants, hospital and clinic assistants, vets and veterinary nurses and is approved for VN training. In addition, it employs education officers, wildlife, farm and animal research experts, press officers, journalists and administrators. For those with an interest in animal welfare but lacking the skills associated with hands-on animal care, there are options well worth considering. The RSPCA runs three wildlife hospitals, one at Taunton, one in Norfolk and one in Cheshire.

Like most charities, the RSPCA welcomes voluntary help in some areas and does employ school-leavers.

At the headquarters in Horsham a spokesman said: 'A lot of young people are perhaps a bit idealistic when they come to us, which can be a good thing if they have their eyes open about the work. Most of them do, but there are stumbling blocks. One of these, of course, is euthanasia. Every year around 70,000 animals have to be destroyed. They are kept for as long as possible, and finding homes for them is our first priority. If there is very little chance of re-homing, if the animal is temperamentally unsuitable, or too old, or – and it's hard to say this – too unattractive, then it has to be put down.'

Hospital and clinic assistants work in the RSPCA hospitals. Applicants must be at least 18 and live near the hospital as the job is non-residential. Training is given for six months on full pay, with instruction given under the supervision of a veterinary surgeon on animal first-aid, euthanasia, clinic hygiene and general reception work.

Kennel assistants are involved with kennel hygiene, animal first-aid, routine exercising, cleaning and grooming. Some aspects of the job are distressing but it can also be very rewarding.

Vacancies for full-time paid employees are limited, as most animal homes are staffed by voluntary workers.

Candidates for inspectors should be aged between 22 and 40, have a good general education, physical fitness and hold a clean driving licence. The five-month training period covers theoretical and field training and includes a final examination. Inspectors must be willing to serve anywhere in England and Wales.

Routine

Whatever their history and whatever their fate, animals in RSPCA care need exactly the same attention as animals in kennels anywhere. At one animal home, in Surrey, a senior staff member said: 'We do get young people coming to work here who want to mollycoddle the dogs, thinking the work here is all grooming and walking and tender loving care. Well, they do need a lot of that, of course. Some of them come in starving and frightened, so they're not like boarding kennel dogs and cats in that sense.

'What I would warn a youngster hoping for this work about is that you do get disillusioned. These beginners are animal lovers and they just don't realise the way people treat their so-called pets. When people come wanting a dog, you really question them. You have to know why they want it and whether they can keep it properly. Sometimes they're looking for status symbols. You do get people in a one-bedroomed flat looking for a Great Dane. You have to be sure the whole family wants the animal, that it will get the exercise it needs, that it will be properly fed. You can tell a lot by the answers people give. And yes, we do often turn people down.

'You've got to do that tactfully, of course. I usually get round it by saying the Inspector will have to call first and see the home. Then I leave it to the Inspector.'

Case Study

Basil is an RSPCA inspector.

'Ever since childhood I wanted to work with animals. There is residential training at the RSPCA College, which is very intensive, dealing with law, lectures from vets, every aspect of the work.

It's hard to generalise about this job because every day is different. One is dealing with animals and people. So much of the work is inspecting complaints of cruelty. Around 100,000 are made each year throughout the country.

Careers working with animals

Most of the complaints are from people genuinely concerned, but 75 per cent of them are unfounded. If an animal is being mistreated or cruelly treated, one feels a controlled anger. You have to feel this anger. There would be no point if you didn't care. But actually there is more ignorance than cruelty. Things are getting worse, with the recession and unemployment. People aren't taking animals to the vet, to avoid fees, and perhaps they cut down on food. It all contributes to neglect.

The thing to remember is that we are not a punitive organisation. We are offering a public service. Most of the time is spent advising and helping. Very few cases of cruelty come before the courts. What is frustrating is that we have no power to remove an animal, however bad the conditions, unless the owner legally signs it over. We can caution people and take statements, then Headquarters takes over, proceeding through the courts if necessary.

People are not always as shocked as you might think to get a visit from an RSPCA inspector. Some are very belligerent and aggressive or arrogant. It is not a job for a timid person. You're faced with heart-breaking options. Yesterday I had to put a boxer dog to sleep. It was nearly blind, very old and lame, and I know it was right, but the fact remains: yesterday I put down a boxer.

We inspect pet shops, all boarding establishments, riding schools and are welcomed in 75 per cent of cases. We have a good relationship with most pet shops, but some are frankly shady – illegally caught finches, that sort of thing. We visit street markets and you need eyes in the back of your head. Deals go on in back streets and we are very conscious of our limited power. We do wear a uniform. I often think we'd do better in jeans with a two-day growth of beard, but we do have a public image.

You need to be a bit of a detective and a bit of a diplomat. You must be able to communicate with people. And you must have a great love and concern for animals and really love the job; otherwise you wouldn't stay. The salary is not high. You need a genuine vocation, and if you've got it, you know you've got it.'

People's Dispensary for Sick Animals

The **PDSA was** founded in 1917 to provide free treatment for animals of people who could not afford private veterinary fees. Today its services are available to pet owners in receipt of means-tested state benefits. The PDSA has a network of veterinary centres from Aberdeen to Plymouth which, together with the PDSA PetAid Service operated through private practices, provides care in some 110 communities throughout Britain.

As Britain's largest veterinary charity it employs around 200 vets, known as veterinary officers, and 240 qualified veterinary nurses. The student veterinary nurse trains over a minimum of two years with two examinations in that period. Once the student has passed the Part Two examination, which consists of a multiple-choice written examination, an oral and practical examination, he or she may apply for inclusion in the RCVS list of veterinary nurses and is entitled to use the letters VN after his/her name.

Salaries

The salaries for PDSA VNs range from £6,042 a year (1996 rates) for trainees aged 16, and older entrants who do not have sufficient qualifications to enrol as trainee VNs, to £11,511 for qualified VNs at the top of the incremental scale.

Other employment opportunities

Increasingly, PDSA veterinary centres employ animal care auxiliaries on varying shift rotas, but mainly at night, caring for in-patients, assisting with out-of-hours emergency treatment, security and cleaning. PDSA veterinary centres also employ part-time receptionists. The work involves dealing with the public in person and by telephone and undertaking associated clerical duties.

The Blue Cross

The Blue Cross is a similar but smaller charity. Like the PDSA it receives no state aid and is totally dependent on donations and legacies. It has three hospitals, three centres providing accommodation for horses under the Horse Protection Scheme, two clinics – and a horse ambulance and two mobile clinics in Dublin and ten adoption centres. The Society employs about 220 staff and the hospitals are approved training centres for VNs.

Today the Horse Protection Scheme often rescues horses from

slaughter and allows them to spend the rest of their lives happily. One man who works with these horses said: 'Sometimes they are brought in by owners who have to part with them but don't want to sell. Some of the horses we have here are old or infirm but we do try, where possible, to put them to work because a horse is happier working. We lend them out, under signed agreement and subject to regular inspection, for things like light hacking or sometimes just keeping another animal company.' The same man operates one of the mobile horse ambulances which attend any events in which horses are taking part, such as gymkhanas or point-to-points, ready to give emergency treatment.

Like the other charities, staff at the Blue Cross have noticed an increase in the number of unwanted animals. A staff member at Kimpton, near Hitchin, said: 'It often happens where not enough thought or family agreement has been given in advance. It's absolutely amazing how many children are supposed to be allergic to dogs or cats. People do make excuses. One chap came in with tears running down his face because he could not afford to keep his dog. It was a really superb Alsatian and they thought the world of each other. But he had just been made redundant. That kind of thing is very sad, and of course no one can explain it to the dog.'

Animal conservation

Animal conservation work can offer close contact with animals that are both amazing and beautiful but much of the work is creating, protecting or maintaining the right sort of habitat. Finding work in this area is not easy. Most people start off by developing their interest in conservation by volunteering. A large number of conservation organisations take on volunteers. Some of these are listed in Chapter 14, but for more information it would be worth obtaining a copy of The Environment Council's excellent *Directory of Environmental Courses 1997–1999*. This lists a number of useful publications, some of which will be relevant to someone considering a career working with animals. All career libraries should have a copy but if not the Environment Council has a limited number for sale.

RSPB

The Royal Society for the Protection of Birds currently runs 121 reserves in Britain. It employs permanent wardens, permanent assistant wardens and in the summer temporary assistant wardens to help with visitors as well as the usual work of the site. Much of the work involves preparing suitable habitats for the birds and physical tasks such as thinning scrub, cutting reeds and clearing ditches are common. Wardens must have excellent ornithological knowledge, being able to identify birds, maintain accurate records and write up regular reports.

From the age of 16 you can apply to become a voluntary warden and duties may include escorting visitors, keeping records of birds and carrying out physical work. From the age of 20 you can apply to be a summer warden. The jobs are usually advertised in the RSPB magazine *Birds* and applicants must be able to work from April to August. Long working hours are par for the course and applicants must enjoy meeting people and be prepared to undertake physical work.

The most outstanding summer wardens might have the opportunity to become permanent assistant wardens or permanent wardens but competition is fierce and demand for places always outstrips supply. In addition, there are a number of opportunities to work at the Society's headquarters at The Lodge in Sandy, Bedfordshire SG19 2DL. For full details of all RSPB jobs write to this address or phone 01767 680551.

Animal sanctuaries

There are a number of animal sanctuaries, some of which offer paid or voluntary work. Hand to Paw's *Animal Rescue Directory* lists most of them, county by county. Sanctuaries usually protect one species or type of animal; these include owls, hedgehogs, guinea pigs, foxes, horses, donkeys, monkeys and oil-spill victims. Each should be contacted individually. The following is one example.

The Monkey Sanctuary

The Monkey Sanctuary was established in 1964 to provide a stable setting in which woolly monkeys, rescued from lives of isolation in zoos or as pets, could live as naturally as possible, in a colony. The care of the monkeys is in the hands of a highly experienced team of people.

The keeper's work includes accompanying the monkeys in the gardens, caring for the monkeys' health and environment and day-to-day well-being, as well as passing on these skills to future team members. When permanent team members are required at the Sanctuary, people are more likely to be selected if they have come to know the monkeys, the team and the lifestyle through being there as volunteers.

Volunteers stay for periods of two to several weeks depending on demand and in exchange for their help they live in with the Sanctuary team. No qualifications are needed but a concern for animal welfare and conservation is. In the summer, work is geared more specifically around receiving visitors, so if specific conservation or maintenance work is more your thing, the autumn and winter is the better time to volunteer. Write to the Monkey Sanctuary for more information.

5 Working with horses

Stable lads

Stable lads, a name given to men and women, start off by performing the most menial tasks. Work starts early with mucking out. Horses are brushed, yards swept, and tack (saddles, harness, bridles) cleaned. Water has to be checked, straw changed and hay fetched. Mash has to be mixed up for three feeds, and other food cleaned and chopped.

A stable lad will also accompany horses to the vet or perform road work, a difficult and hazardous task which can result in riders falling off. To move on from basic work or to be trusted to ride the horses most prized by the stables, lads will have to prove themselves to the head lad, who makes all the day-to-day decisions. Progress is gradual, standards strict and at the end of years of hard work, few become successful jockeys.

Training for stable lads is given by private trainers and at the British Racing School and the Northern Racing School.

A list of trainers (over 600 in the British Isles) is given in *Horses in Training*, which can be bought from bookshops and is also in most libraries. Job applications can be made direct to a trainer and for those with experience of riding and looking after horses it is worth contacting the National Trainers Federation which distributes a list of prospective stable staff to trainers when they have a vacancy in their yard. *Sporting Life*, *Racing Post* and *Horse and Hound* may also carry job advertisements. Many trainers take on school leavers on Youth Training and Modern Apprenticeship programmes. Experience is necessary before trying for work in a racing yard as trainers do not normally have time to teach complete beginners. To work as a stable lad in flat-racing yards apprentices should weigh less than 9 stone (51–57 kg).

Previous experience is not essential before applying for the courses at the two racing schools.

The training programme is on the lines of a real racing stable with instruction on riding, grooming, feeding, mucking out and lectures on the parts of a horse, its health, possible ailments, lameness etc. The courses are open to 16–19-year-olds and satisfactory trainees are guaranteed jobs in racing yards. Trainees should have good eyesight without glasses, no chest complaints or asthma. Good exam results are not essential but applicants must be alert, quick to learn and able to read and write. Training, accommodation and food are free.

The courses at the schools lead to National Vocational Qualifications (NVQs) in Racehorse Care and Management up to level 3.

A few stable lad trainees do go on to become jockeys, but out of 4,500 stable workers, only 200 are professional jockeys; those who do not can become head lad, or a travelling head lad or assistant trainer, or even eventually make it to trainer.

Would-be flat race jockeys are apprenticed to a trainer for a period between the ages of 16 and 20, working as stable lads and riding in races if they are good enough. Steeplechase, or 'jump', jockeys do not have any formal training, but are apprenticed as 'conditional jockeys' up to the age of 25, with instruction being given by the trainer. There is also one course each year for apprentice jockeys at the British Racing School. Sixteen-year-olds who want to be jockeys on the flat must not weigh more than 7–8 stone (44–51 kg); the limit for jump jockeys is around 8–9 stone (51–57 kg). Often jump jockeys are ex-flat race jockeys who have grown too heavy, or who have been riding previously in point-to-point races.

For more information, contact the Racing and Thoroughbred Breeding Training Board (RTBTB).

Pay

Low pay is legendary. The minimum rate of pay for a 40-hour week, including overtime, varies between £82.89 for a lad aged 16 to £164.00 for a lad who has completed NVQ level 2 with 18 months' experience in a racing yard.

It takes courage to complain about pay or conditions, knowing that there are dozens waiting to take over if you don't want the job. To fight for better standards a union, called the National Association of Grooms (NAG), was started by Christine Stafford-Hughes, who said: 'Too many stable owners run their businesses on the basis that there will always be a succession of starry-eyed girls who will do absolutely anything to work with horses.' Conditions are improving, but there are still too many cases of exploitation. NAG has now become the Equestrian Section of the Agricultural Trade Group of the TGWU.

Stud work

The stud year has three separate phases. The busiest time for the stable staff is from February to June, which is the breeding season. The months from July to October are less busy, with the most important work being to prepare the yearlings for the autumn sales. November to January is the time for routine maintenance and for taking holidays.

Stud workers are paid according to experience; wages are normally higher than the minimum agricultural wage. Opportunities for promotion are limited, but the positions of stallion men or stud grooms carry higher salaries for extra responsibilities.

Formal qualifications are not required but some experience of handling thoroughbred horses is preferred, together with some knowledge of stable management and general horse care. Someone under 19 years of age can enter the industry as a trainee stud hand via the National Stud's introductory course. Alternatively, they can write; to the Thoroughbred Breeders Association which has a list of studs willing to take on trainee stud hands; to studs (their addresses can be found in the *Directory of Turf*); or use the situations vacant columns in the racing papers.

Stud workers usually work towards NVQ qualifications in Racehorse Care and Management while they are employed but it is possible to take a college based vocational course such as the highly regarded one- and two-year courses in Stud and Stable Husbandry offered at the West Oxfordshire College of Witney and The National Stud's intensive National Stud Diploma. The Thoroughbred Breeders Association will send out a list of horse-related courses and the colleges which offer them if they receive a large sae with 38p in stamps.

The National Pony Society offers the Stud Assistant's Certificate and the NPS Diploma (non-riding) in Pony Mastership and Breeding. The Society publishes a list of training establishments in stud work; trainees can take the NVQ in Horse Care and Management levels 1, 2 and 3.

Show jumping

For many show jumping fans, the peak of ambition is to act as groom to one of the competitors. While many treat grooms well and pay them fairly, this is an area of great exploitation. One regular eventer said: 'Frankly, this is an expensive business and I'm sorry to say there are people who exploit these youngsters. It's regarded as one of the little economies. There are so many keen youngsters, queuing up for jobs, hanging around at events. They want to get into the world of eventing and show jumping and I suppose this is one way to do it.'

Case Study

Lisa *is a groom and qualified instructor.*

'You can find yourself sleeping in the stockman's tent, huts, all sorts of places, travelling all over Britain. Mind you, you can get it cushy. But I wouldn't recommend it to someone young. You've got to be a very good rider and know all there is to know about looking after a horse. Some eventers won't let anyone exercise their horse. But many do, and you can come off. That's when you discover your employer doesn't believe in buying insurance stamps – a not uncommon occurrence, believe me.'

Working with horses

Riding schools

Every year hundreds seek jobs in riding schools for a year (September to September), hoping to become instructors. If they can afford it they pay for intensive instruction. Talland School of Equitation in Cirencester is one of the riding schools offering this sort of training. Approved by the British Horse Society, it offers a variety of options for trainees, lasting from 17 weeks to a year. All prices include full-board accommodation and tuition but trainees are expected to do stable work as well. Students pay £241 a week for 17 weeks, £205 a week for 24 weeks or £95 a week for a year. For details of other courses contact the BHS. The alternative is to work in return for tuition and coaching for exams. Here, too, there are dangers for the inexperienced. Some find themselves grossly overworked – even some who are paying for tuition – and badly taught. It is perfectly possible to end up with no qualifications, worn out and unable to go on to the instructing job hoped for.

Case Study

Angela runs a riding school in Suffolk. Her stables are approved by the British Horse Society, which carries out regular inspections of all its listed establishments.

'My advice to anyone looking for work in stables is to write to the Society. It seems an obvious first step, yet so many people don't do it. There is a lot of ignorance, or innocence, on the part of both youngsters and their parents. I advise them to have a written contract with their employer, setting out hours of work, time off and conditions. It should be read by parents, with a solicitor's advice if necessary. And everyone should find out if there will be insurance cover. Quite often there will not. I know of one case where a girl came off a horse very badly. The fall was entirely due to a broken bridle. She was laid up for weeks without pay. And when she returned, the owner charged her for the broken bridle.'

It is not unusual for people to find after starting that they do not want to become instructors after all. If they have learned enough about horse care, including treatment of minor ailments, stable management and routine, and have good riding ability, they can still achieve qualifications and find satisfying work in stables. They can sit for the BHS examinations (see below) or the Association of British Riding Schools' Assistant Groom's Certificate and Groom's Diploma.

British Horse Society (BHS)

The BHS holds examinations at centres throughout the country, to give qualifications for a career working with horses. There is a comprehensive syllabus that includes horse care as well as riding.

Many options are open, especially for those who intend to look after horses rather than ride them. It is possible to climb up a ladder of advanced qualifications but all trainees start by learning the basics. The first stage, the British Horse Society Assistant Instructor (BHSAI) Certificate, is offered under Youth Training but the more advanced qualifications would be taken after 18. These are the Instructor's Certificate, the Stable Manager's Certificate and the Fellowship of the British Horse Society. The BHSAI course includes components on horse physiology and psychology, health, management, general handling, feeding, sick horses and general knowledge subjects such as insurance. Syllabuses for all the BHS examinations are available from the BHS (20p each plus sae).

The BHS produces a book, *Where to Ride*, which gives information on riding establishments which are approved as places of training. It is available from large bookshops or from the BHS Approvals Office (£6.95 including postage and packing). Most of the approved establishments that offer careers training can offer working student or fee-paying places.

Courses

A great many courses exist at colleges throughout the country (see pages 76–81). Contact the British Horse Society for further details of examinations and courses.

National Vocational Qualifications/Scottish Vocational Qualifications

The training body of the horse industry, the National Horse Education and Training Company, is accredited under the new National Vocational Qualification scheme to award the NVQ levels 1, 2 and 3. The Certificate is awarded on the basis of assessment of competence in core units.

Core units are: Health and Safety at Work with Horses; Basic Handling of Horses; Watering and Feeding Stabled Horses; Stable Routine; Grooming and Care of the Stabled Horse; Basic Care of Horses at Grass; Horse Clothing; Saddlery/Harness; Horse Health; Transporting the Horse; Care of the Horse's Foot; Lungeing/Long Reining. Optional Units are: Equitation; The Racing Industry; Heavy Horse Work; Breeding.

Normally, assessments are designed to be work-based at the stables or at the centre providing instruction; assessors are the college teaching staff, riding school instructors or senior members of staff of other equestrian establishments.

Candidates must enter through an approved equestrian centre or local agricultural college. There is no age limit for NVQ awards.

Modern Apprenticeship

The horse industry has recently introduced its first recognised Modern Apprenticeship scheme. This is likely to become *the* work-based training programme for future senior equine staff. All apprentices receive a basic wage and access to free NVQ training. In return they commit themselves to work for their employer for the duration of the apprenticeship – usually three years. For further information contact the National Horse Education and Training Company.

The Association of British Riding Schools (ABRS)

The ABRS has a career and examination structure for grooms starting with the ABRS Preliminary Horse Care and Riding Certificate which is offered to YT trainees and other young persons involved in a career with horses. The ABRS Assistant Groom's Certificate is intended to provide a guide to employers in the selection of competent grooms. The holders, while capable of working on their own for a limited time, still require some supervision. The ABRS Groom's Diploma indicates that the holder is a competent groom who is capable not only of working on his or her own, but of organising the work of other staff. An experienced diploma holder is considered to be stable manager material. No GCSE passes are required for any of the ABRS examinations. Details of syllabuses can be obtained from the ABRS secretary, and the Association has a list of approved establishments which can offer training for their examinations. A list of establishments is also available from the British Horse Society.

Farriers

Stables could not manage without farriers to shoe the horses, and although people sometimes think of this as a dying craft there are, fortunately, still plenty of them about. Many farriers work with the Household Cavalry, the King's Troop Royal Horse Artillery, the Royal Army Veterinary Corps or the police, but it is still possible to be apprenticed to one who deals with civilian horses. A blacksmith is not a farrier, but a person who works in metal, usually wrought iron. And the second myth to be dispelled is that people shoeing horses have to be brawny and burly. There are many women farriers.

Apprenticeships

A farrier cannot practise without being registered by the Farriers Registration Council. Qualification to register is obtained by:

1. successful completion of an apprenticeship with an Approved Training Farrier;
2. passing the examination for the Diploma of the Worshipful Company of Farriers.

Army personnel follow a similar training but must pass the Army Trade Test BII in farriery to qualify.

Candidates for apprenticeships in farriery must be at least 16. They will be asked to take a medical examination (at their own expense) which must include a test for colour blindness, although this is not a hindrance to applicants.

The minimum educational requirements for acceptance on to the Farriery Apprenticeship Scheme are four GCSEs at Grade C (or the equivalent qualifications for Scotland, Northern Ireland and the EC), one of which must be English language or equivalent.

For further details contact the Farriery Training Service (FTS). If you want to obtain a list of approved training farriers (there are about 250 in Britain), the Farriers' Registration Council will send you one on receipt of a large sae. There may be some grant aid available through Youth Training and the FTS so it is worth asking. Apprentices aged 21 and over may be eligible for local education authority support. But there are few grants and farriers may have to bear the brunt of the £6,500 or so needed to complete the four-year training programme.

Pay

There is no set wage for qualified farriers, and income depends to a large extent on competition, quality of work, and the amount of work the individual farrier is able to command. For apprentice rates, contact the Farriery Training Service.

The armed services

The King's Troop Royal Horse Artillery

The Troop is seen every summer pulling the guns during the musical drive in the Royal Tournament. It is also responsible for the firing of royal salutes in Hyde Park in London on special occasions, such as royal anniversaries and state ceremonies. The Artillery is the branch of the army responsible for guns and missiles.

New recruits do not have to be able to ride. There are Monday to Friday courses every month (except during the Royal Tournament in July) when about six potential recruits are given the chance to find out if they would be suitable; the course includes riding instruction. If they are interested in joining the Troop, they must then go through normal army selection procedures. Those who are not skilled enough to ride on parade and in the Royal Tournament may become limber gunners or stablemen responsible for the care of the horses. All get a chance to ride out on exercise and each man has his own horses in his charge in the stables. Members of the Troop include farriers and saddlers, and everyone learns to ride – including tailors, storemen, vehicle drivers and batmen.

Mounted dutymen in the Household Cavalry

The Household Cavalry is an armoured regiment using tanks whose soldiers may be crewmen, drivers or mounted dutymen. They provide the Queen's life guard at Horse Guards Parade and carry out duties for the royal family and visiting heads of state. Like the King's Troop Royal Horse Artillery, the regiment includes farriers and saddlers.

Royal Army Veterinary Corps

The Royal Army Veterinary Corps (RAVC) employs veterinary surgeons, who help to choose horses and dogs used by the army and are responsible for their care; also veterinary assistants, farriers and dog trainers.

Full details of careers in the armed services are available from local Army Careers Information Offices. All recruits must go through the normal army recruitment procedure.

Mounted police

Horses in the police service have a far greater role than those used for ceremonial duties in the armed services. Riders are very skilled and horse and rider play a vital role at events where there are large crowds, such as football matches, race meetings and demonstra-

Working with horses

tions. Not all forces have a mounted section, and in those that do the branch is usually quite small and the opportunities for promotion are limited.

An officer must complete two years' foot duty before applying for a post in the mounted branch to ensure that he or she is well grounded in general police work. The majority of recruits have no previous experience of working with horses, and are taught to ride and about the care of the horse and its equipment. On a normal day a mounted officer patrols for three hours, the rest of his eight-hour tour of duty being spent on grooming, general stable work and cleaning tack.

For further information contact the recruitment officer of your chosen police force or the Home Office which issues a list of forces which have a mounted branch. Information from: the Police Department, Room 514, Home Office, 50 Queen Anne's Gate, London SW1H 9AT (for the police service in England and Wales) or Police Division, Scottish Home and Health Department, St Andrew's House, Edinburgh EH1 3DE (for the police service in Scotland).

6 Working with dogs

Dogs are quite often not only man's best friend, but man's lifeline, living or protector. The various occupations in this chapter include guide dog trainer, police dog handler, kennel hand and general dog trainer.

Kennel work

Although there are a great many private training courses for all aspects of dog care, most kennel owners prefer to employ people straight from school. One owner said he likes people to come to work for him before they have softened up in an indoor job. The work is largely outdoors, starts early in the morning and is physically very demanding.

Kennel work is not particularly well paid and varies, depending on the region you live in and the type of accommodation arrangement that goes with the job. This can range from a room in a house, to a flat or a caravan. Quite often food is provided and junior kennel staff may live as part of the family unit, sharing cooking and eating arrangements.

There is no minimum wage and even staff who live out may earn as little as £40 a week. Mostly though, the wages are better than this and people with an NVQ level 2 qualification should expect to earn between £110 and £130 per week if they live out and less if accommodation is provided. Senior staff members with greater responsibility can expect around £200 a week. Even in the best-run establishments, where staff have proper working hours and adequate time off, the work is exhausting. However, skilled staff are in short supply, so those in their early twenties often find themselves with responsible jobs in management positions.

Case Study

John Hyde owns the Star Boarding Kennels in Chessington.

'They come along, full of willingness and spirit, these young beginners, and for ten days they keep up the pace. Then they either pack it in, or slow right down. They wander about in a daze, knocked out by sheer hard work.

That period can last for anything up to three weeks. Our senior staff are very experienced and they know the symptoms. They are pretty understanding. If the beginners come through that, they'll do. In my experience only about three or four out of, say, 30 do come through it. They're usually girls.

Anyone who comes along thinking they'll spend their time stroking lovely little cats and dogs all day has the wrong idea. Very few who come here for a job have any notion of the amount of work involved. Seventy per cent of it involves keeping the animals absolutely immaculate, cleaning bedding and runs, disinfecting and grooming. It starts at 7 am and the first break in the day comes at 11 am or 12. It's a five-day week, but that includes weekends and Christmas, which means shift work. You can't just fold animals up and put them in a filing cabinet at weekends. But we don't work our staff to death as some kennels do.'

Standards

Like all owners, Mr Hyde is ready to concede that there are a lot of badly run kennels. Some are too inefficient to provide much in the way of training for a beginner. Some are understaffed and ask a great deal too much from the workers they have.

How can an outsider looking for a job tell the good from the bad? One short, perfectly serious answer we were given to that question was: 'Use your nose.' A well-run kennels smells clean. A badly run one usually does not. The dogs themselves are the best advertisement for a good establishment. It does not take an expert to tell the difference between lively, properly fed creatures who are exercised regularly, and depressed, badly fed dogs who spend much too much time in unhygienic cages.

Training

Training is given on the job. Trainees can combine practical work with a correspondence course to gain the National Small Animal Care Certificate. The course is run by the Animal Care College and is designed for trainees at boarding, quarantine, breeding, rescue and sanctuary kennels and is adaptable for use at pet shops and grooming establishments. The NVQ in Kennel Supervision, level 3, is awarded through on-the-job assessment.

Case Study

Lynn is 24 and a senior kennel hand. She has been working in kennels since she was 14.

'I used to come here at holidays and weekends. You couldn't keep me away. The pay was low but that didn't bother me. It was the dogs I was interested in. I used to clean out kennels, walk the boarders, learn to groom, and I couldn't wait to come back each time. It never occurred to me there might be a dog I couldn't get on with. I do love dogs, but I'm not sentimental about them. I try to understand the way they think, what's going on in those furry minds. You have to, believe me. Actually, I have noticed a change in dogs over the years. Maybe it's due to inbreeding. Maybe people are bringing them up the wrong way. But years ago you never saw a neurotic retriever.

You start as a kennel maid, cleaning out, answering the telephone, looking after the puppies. I still clean out. Everyone has to do that and it can come as a shock to beginners. You show them how to do it, make sure they're doing it properly, teach them the rules. Then you should be able to trust them to get on with it. But I've got a little lad here, and he needs watching. He'll skimp the job if he thinks he can get away with it. To be honest, I really can't understand that. If you like dogs and you want to work with them you've got to ask yourself, now what can I do to make this dog feel comfortable? Is the bed clean and dry? Is the water bowl clean? Is it empty? You don't do it because there's someone on your tail, you do it from affection for animals. And if you've got that, you won't think of the mucky side as boring.'

Greyhound racing

A kennel worker in greyhound racing has a different set of responsibilities. The daily routine revolves around the general welfare and training of the greyhound but at the racecourse the kennel hand guides his or her greyhound through the pre-race procedures, the weighing, veterinary examination and parading on the track. In the absence of the trainer the kennel hand is also responsible for the security and safety of the greyhound. Although the work is hard and demanding, it can also be hectic and exciting, especially if a high-class racing greyhound is placed in your care. There is no formal career path but those with ability can assume professional trainer status after two years' experience as a kennel hand. The National Greyhound Racing Club can provide a list of licensed racecourses.

Dog training

Training usually starts off as a part-time interest. Those who consider it their vocation are the ones who have the dedication and determination to turn their hobby into a full-time job. It is also possible to get work experience before becoming a full-time trainee trainer, as one owner of a dog training centre explained:

Case Study

Brian is the owner of a dog training centre.

'I prefer to take on school-leavers here. Usually they come to us while still at school, for a week. At the end of that, if they are interested and we find their work satisfactory, we suggest they come at weekends. They are fully paid for any work they do during that time. I do not believe in making a profit out of someone's enthusiasm.

Beginners are taught every aspect of kennel work, maintenance and dog care. It is physically demanding work, but everyone here shares it. We encourage animal husbandry training, giving the learners the chance to work with a veterinary surgery to increase their experience. Actually, it works in reverse too. Vets send their nurses to us during training. They begin dog training after a few months. This is a skill which can be acquired. The unskilled trainee does, after all, represent the general public who own these dogs and to whom these dogs will return. They need to have assurance with the dogs, and that too can be acquired.'

Many trainers work alone, specialising in training gun dogs, greyhounds, guard dogs or domestic pets. They may have around ten dogs at a time on their premises for training. Private companies are now taking over the work of training police and military dog handlers and dogs, in searching out drugs, arms, explosives and humans. Some of these handlers are sent overseas to tackle jobs formerly dealt with by the military; training is restricted to people with a service background, and trainees are first assessed to ensure that they have a real affinity with animals.

It is worth remembering that unless you choose one of the public sector employees, only the Guide Dogs for the Blind Association has a recognised training programme. In the private sector there are no NVQs or any other structured training programmes and no state aid. The British Institute of Professional Dog Trainers call dog training a do-it-yourself profession. Ask them for their useful careers leaflet and details of their Certificates in Instructional Techniques.

Case Study

Ann *is 24 and a highly skilled trainer.*

'On my Easter holidays, just before I left school, I was handed a cross Labrador and told: 'If you can train this dog, we'll take you on as a trainee trainer.' I've never worked so hard at anything, and I did it. It's true anyone can be taught to do it, but some will never make trainers and you can tell. Equally, people who come as naturals are few and far between. You need a feeling for dogs and an understanding of them. You've got to make it clear to the dog who is in charge, but there's no point in frightening him. Take this one. She's inclined to snap, particularly when she's with her owner. She's come here to learn better manners because she had a go at a builder. Her owner thinks the dog is over-protective. Well, I think she's timid. She wants her confidence building up. She might be

Working with dogs

growling or snapping, but her ears are down and look at her tail. Right between the legs.

We had one real delinquent here – a Yorkie that bit people. His owner wrote in despair to a very famous dog expert who wrote back advising him to take all the dog's teeth out. Well, we laughed but of course it's not really funny. We told him we thought that was a bit drastic. We use a check chain, give a sharp jerk, tell the dog off, then – when he's thought it over – a word of reassurance.

We work the dogs here in the yard, simple commands like 'sit', 'heel', 'down'. Then we take them out. This nervous one I shall take to shopping centres, into crowds, to get her used to it. You've got to remember you're not training these dogs for yourself but for other people.

It is a pleasure to take an untrained dog, or a badly trained one, which is worse, and work patiently with it and see it turn into an obedient, sensible animal.'

Police dogs

Police dogs attract a lot of attention. There is something fascinating about the fact that they look perfectly ordinary while everyone knows they are one. As one policeman describing an off-duty incident put it: 'I was in plain clothes at the time, and so was my dog... heavily disguised as a family pet.' Like all good disguises, that one was based on reality. Police dogs are family dogs, living with their handlers from puppyhood. Their handlers are normally married, so there is someone at home to look after the dog, and they must have a garden.

Training

At eight weeks a puppy is allocated to a handler, who takes it home and treats it as any other owner would, training it in simple obedience and building up a strong relationship with it. At nine months old the puppy and handler go to Keston, Kent, to the Metropolitan Police Dog Training Establishment for a week's obedience and nose work.

When the dog is a year old it goes with its handler on a 12–14-week course, and this is the biggest step in its life. Some dogs don't qualify, which is sad for everyone, particularly the handler who has probably waited for two years to be allocated a dog. On this training, the dog learns to follow a scent, search out criminals and property, and chase and hold suspects; it is also taught never to bite. Dog and handler have to attend frequent refresher courses at one of five training centres throughout the dog's working life, which lasts about seven years. A retired dog usually remains with its handler who is then given another puppy.

A very senior policeman who spent most of his life working with dogs said: 'First you've got to want to be a copper; don't forget – there are more people who want to work with dogs than there are dogs.'

It's not possible to join the police as a handler. Two years on the beat must be completed first. Dog handlers and their dogs now have to be licensed annually.

Guide dogs

Another specialised branch of training is carried out by the Guide Dogs for the Blind Association. There are seven centres – at Leamington Spa, Exeter, Bolton, Forfar, Middlesbrough, Wokingham, Redbridge and the Breeding Centre at Bishops Tachbrook, near Leamington. Each centre hires its own staff. Promotion to training assistant is usually made within the Association, so anyone hoping for this work would do well to start by finding a job as a member of the kennel staff.

Kennel staff

This is more varied than many kennels jobs. It includes the usual grooming, feeding and exercising routines, but for at least four periods of the year it involves contact with future owners of the dogs. Making them welcome during their training stay at the centre, morale-boosting and practical help are part of kennel staff duties, so obviously quite a special person is looked for by the controller of a training centre.

Applicants for kennel staff must be at least 18 years old, with at least two GCSE/SCE (or equivalent) qualifications in English language and maths. Kennel staff gain City and Guilds Kennel Staff qualifications and can eventually take the C&G Kennel Supervisor qualification.

Kennel staff live in at the centre; food and accommodation are provided free of charge.

Guide dog trainers

Applicants for guide dog trainers should be at least 18 with three GCSE/SCE passes, including English language, maths and preferably a science subject. They should also have domestic or work experience with animals and work or social experience with adults. A driving licence is essential. The training leads to the City and Guilds Guide Dog Trainer qualification.

The trainers provide the dogs with the real introduction to their life's work. They take over a nine-month-old dog who has lived with a puppy walker family and ease it into its new routine.

They begin simply, taking the dog for pleasure walks, assessing it all the time. After three weeks the dog is taught to walk at a comfortable speed, how to respond to the harness it will wear as a working dog, and how to concentrate on the voice of the training assistant.

After eight weeks an obedient, responsive dog is ready to be handed over to the next trainer, a mobility instructor.

Free food and accommodation are provided for trainers during the first six months, which is a compulsory living-in period.

Mobility instructor

There is a three-year training course to become a mobility instructor, with an examination at the end of it. The course covers dog welfare and psychology, selection and training and aspects of veterinary care, and trainees must pass City and Guilds examinations. The trainer is also instructed in the psychology of blind people and the practice training of future owners in the use of a guide dog.

The trainer has to teach the dog how to judge height and

width. It must learn that some occasions call for disobedience, such as ignoring a command to go forward into a situation it can see is dangerous. The dog is taught to regard the human at the end of its harness as an extension of itself and allow for that when it makes basic decisions, such as whether or not a space is wide enough to pass through or whether there is headroom. Naturally, this highly specialised training calls for a lot of affection from the instructor and desire to please on the part of the dog. One of the most difficult parts of the training is helping the dog transfer its loyalty to its new owner. Mobility instructors visit the dogs in their new homes during the first few days, travelling regular routes with them and giving advice to the owners.

Instructors receive free food and accommodation during the first six months, which is a compulsory living-in period.

This is a demanding and stressful job and applicants must be at least 18 years of age. Qualifications needed are five GCSE or equivalent passes including English language, maths, a science subject and a social science subject. Work experience should include dealing with groups of people from various social backgrounds, preferably including the disabled. A driving licence is essential.

Salaries

Starting salaries for guide dog trainers are £8,591 per annum, rising to £13,246 after training. Salaries for mobility instructors start at £12,615 and rise to a maximum of £18,885.

Hearing Dogs for the Deaf

Although a comparatively young organisation, with fewer job opportunities than at Guide Dogs for the Blind, Hearing Dogs for the Deaf offers unique opportunities for people wanting to train dogs to understand and react to certain noises, such as a telephone, a door bell and so on.

Candidates are expected to have an outgoing personality, an interest in dog behaviour and experience in training and welfare.

Membership of dog clubs, joining agility and obedience classes or training your own dog to a specialist level would all count. Registering your dog as a PAT (Pets as Therapy) dog and paying regular visits to sick and lonely people will also demonstrate interest and commitment to the concept of dogs helping people.

Previous applicants have worked in veterinary surgeries, or have worked professionally at a kennel; others are graduates with relevant degrees in zoology, psychology and animal science and some have worked for Guide Dogs for the Blind. You should have a good standard of education, with some GSCEs, and a full driving licence.

Guard dogs

Training guard dogs is a job calling for fine judgement. There is no point in turning an animal into a potential killer. It has to be taught the difference between friend and foe, how to scare off intruders, how to deal with aggression and how to defend itself and its owner's property without inflicting serious damage.

Someone has to act the part of the bad guy during training; this is not always a popular role among young trainers, whose instinct is to have a good relationship with the dogs.

Case Study

Clive is 16 and has begun training guard dogs.

'I do like dogs and I've always wanted to work with them, but guard dog work doesn't worry me, in fact I like it. I'd no idea if I would be good with dogs and I was a bit nervous at first. I had to start by cleaning kennels and so on, which I'm not too keen on. What I really like is training. I bait the dogs, with a stick in a sack. No, I don't think it strange for a dog lover to do that kind of thing. I enjoy making them angry. Dogs enjoy guard work. I can see why some people don't like it. The person who baits a dog has to keep out of its way after that and it really worries me that I might forget. The dog won't forget. But I'm never afraid of a dog otherwise.'

7 Pets

Pet shops

Some young people who want to work with animals make their own opportunities. Not everyone can set up his or her own business, although there are still surprising numbers of people who do, usually beginning by working for other people until they have the expertise.

And not everyone who begins by taking holiday jobs in a pet shop ends up managing a chain. But it can be done. Neil Fox is someone who did it at 23. He runs a pet shop near the Angel in Islington, London, and is area manager, with responsibility for several other shops.

Case Study

Jane *is a pet shop manager.*

'It started when I was still at school and I had a fish tank. I used to come in here to buy things for it. I just liked animals so I came around a lot, and when I left school I asked for a job.

The day starts at 8 am, before the shop opens to the public. I let the animals out while I clean and disinfect the cages, and I let the puppies run about the shop, for exercise. Of course it's chaos, but they love it. They make a mess, but you have to tolerate that. Where else can they make a mess? Cleaning up is all part of the job.

You do spend most of your time selling from the shelves, but every time you sell an animal you get to know the owner. They come back for food and sometimes advice, and let you know how the animal is getting on. I can tell who will give an animal a good home. They usually come in to see it once or twice before making up their minds. Then they bring the family, or the girl friend. If I think they wouldn't look after it properly, I wouldn't sell the animal. I can just tell. Maybe I'm making a mistake, but I'd rather do that than sell an animal to a buyer who would not look after it. I just tell them it's sold.'

Fifteen thousand people work in pet shops in Britain. Duties are such that a high degree of dedication is a primary requirement. Staff working in a pet shop need to develop their knowledge of livestock husbandry, products and customer care. The only recognised qualification for pet shop employees is the City and Guilds Pet Store Management Course 7760. This is available from some colleges or by correspondence through the National Extension College and the Pet Care Trust.

Animal grooming

The most usual business for those who want to set up on their own is animal grooming. As anyone who has ever forced so much as a Chihuahua into a tub knows, dogs don't take joyfully to this process. And there is a great deal more to it than a canine shampoo and set.

Animal groomers work with dogs, and are also known as dog or canine beauticians, or simply as dog trimmers or clippers. Fashions have changed since the days of elaborately shaped and beribboned French Poodles, but owners are still prepared to pay well for a grooming and trimming service.

The grooming routine includes bathing the dog, drying it (probably with a hairdrier), brushing and combing the coat, trimming the nails, cleaning teeth and ears and dealing with fleas or other parasites, if necessary. The final result is a pleasant-smelling dog with a gleaming coat.

The trimming part of the process is particularly important for long-haired dogs – especially in the summer. They may need to be clipped every six weeks so that the coat stays in shape; but during the winter the coat is normally left to grow long, so there is less work at this time.

Dog beauticians may be working for private pet owners, or for those who want to show their dogs. Although they may work for a shop – or 'parlour' – they can also work from home or within boarding or breeding kennels, a pet shop or vet's surgery. Working from home can mean dealing with the dogs on your own premises, or travelling to the owner's home or kennels and working there.

As well as being able to handle dogs, even those that do not like being bathed, a dog beautician also has to be fit enough to spend a great deal of the day standing up. It is not a job for people with skin allergies.

Training is usually by apprenticeship with an established pet parlour or beautician, or within kennels, where the training would be included as part of general kennel work. It is important to make sure at the outset that the training will be of a good standard; it will take from one to three years. The same cautious approach should be taken with the private fee-paying courses, though most are good. These may last only one week, or a year; the longer ones should include instruction in the different styles used for the various types of dog, aspects of dog care, and practical experience.

Applicants may also secure a place on a Youth Training programme. All training should be delivered by qualified trainers who hold the City and Guilds Dog Grooming Certificate 7750, and who preferably have achieved or are working towards the Advanced Grooming Diploma.

All dog beauticians are advised to take these qualifications as they show potential employers and customers that you have reached a qualified level of expertise. A careers leaflet is published by the British Dog Groomers Association.

Case Study

Peter *served an apprenticeship in the grooming section of a large store and has recently opened his own establishment.*

'I wouldn't say canine beautician is the best thing to tell people you do for a living, although that is the correct term. You do get some funny reactions. I get some remarks about the name of my shop too – Peter's Posh Pets. Actually, the reason for that is quite basic. I had to get the name Peter in, because previous clients knew it, and I tried to make it memorable. PPP seemed not entirely inappropriate.

You train on the job as an apprentice – no more nerve-racking than training as a hairdresser. Less, I'd imagine. You trim a cocker spaniel a millimetre shorter on the right and he's not going to hit you with his handbag.

I wasn't interested in training school. How do I know they're qualified to qualify me? If you do this job properly, and I do, you don't just shampoo the dog and push it under the drier. You examine its teeth, clean them if necessary, make sure they're sound. You attend to their ears and nails and, when they are in the bath, express their anal glands – not pleasant, but quickly done.

It's pretty hard work, and dirty work as well. It is usually done by girls, but I find it an advantage to have a deep voice and a forceful personality when I take on the bigger dogs. I don't have much trouble with them, but I'll muzzle a dog if it's got ideas. There's no point in being lax with them. They're like little kids. They need firm handling. I'm never afraid of a dog, although I've certainly been bitten. The minute you're afraid, the dog has won.

This can be a seasonal job for some, although not for me because I'm uptown. Work can fall off a lot during the wet and cold season and anyone taking it up should know that. It's not well paid, frankly. I pay myself only a modest salary and there are times when all the headaches involved in running a shop make it seem hardly worth while.'

Animal photography

With imagination, ingenuity and a bit of lateral thinking, skills won in other areas can be put to use in animal work. At one end of the scale, an animal-mad baby-sitter of 17, suspended between school and college, suddenly realised she could make at least as much money and have at least as much fun dog-walking.

A considerably larger venture was undertaken by Sally Anne

Thompson. She was determined to become a photographer, in spite of the attempts of her father, who had a studio, to put her off. He felt it was an overcrowded profession. Sally Anne persisted and created a whole new profession by combining her skill in photography with her love of animals.

Case Study

Jim is an animal photographer.

'With some photographers of animals, love of the animal comes first and photography second. I think it has to be the other way round. You must have the commercial training.

I was lucky enough to get a job as darkroom assistant and went to night school at a polytechnic. I photographed the occasional dog, and zoo animals, then I had one of those fortunate encounters with someone who bred dogs and who encouraged me. There is a market among people who show dogs. They want photographs to send to potential buyers or use in trade publications.

I wouldn't say you need a rapport with animals so much as some sort of idea of what a picture should look like. It's hard work too. You take a shot and people say, 'That's it'. They're amazed if you tell them it will take a couple of hours.

Some dogs are nervous, and some are awkward to start with. You have to work out what makes them tick.

Most of the show dogs, if they are worth photographing, are pretty sensible. They are used to the show ring. I photograph horses too – of course I don't take them into the studio. In fact, in almost all cases, even pet studies, I prefer to go to someone's home. It can take a day, and another day processing. Animal photographers might charge anything between £100 (or less if they were just starting out) and £1,000 per day.

You can sell animal pictures to so many different places. I do a lot of work for books. It wouldn't cost that much today to set up, with so many good cheap cameras, but you have to be exceptional to get very far. You could go to show jumping or eventing and send proofs to the competitors. They might be so thrilled that they will order the photographs. But it's best to start by doing it at weekends, with a guaranteed job during the week. It's also essential to get good advice on business management and accountancy. You must be prepared for competition and you really have got to have something exceptional to offer.'

8 Zoos

Introduction

Zoos have come a long way since naturalist and explorer Sir Stamford Raffles established the Zoological Society of London in 1826. The world's first zoo, a small area of Regent's Park, housed animals with little consideration of their habitat or diet. Throughout Victoria's reign in the early part of this century, the growing collection of animals was kept in badly constructed cages and concrete pits.

Even as late as 1968 members of the public were still able to feed the animals with left-over stale bread and sticky buns from their picnics. With 15,000 to 20,000 visitors a day in the height of summer it is easy to imagine the dietary effect.

Things are quite different now. Diet has become much more important. One of the daily duties of the keeper is to prepare a range of foodstuffs for the animals. As each species requires a slightly different diet, getting the balance right is crucial.

Habitats are more considerate of animal needs and some zoos have pioneered in-captivity breeding programmes to help re-establish a number of species in the wild.

The Federation of Zoological Gardens of Great Britain and Ireland estimates that 20 per cent of zoos are already participating in global conservation strategies for endangered species, offering some wild species the only guaranteed chance of survival. Zoos are not an alternative to habitat conservation, but they are an essential safety net for the most endangered species.

As conservation work increases, zoos will employ specialist zoologists, conservation strategists and ecologists. A degree is a prerequisite for this sort of work.

Zoo keeping

Zoo keeping attracts thousands of applicants each year. The biggest employers are London Zoo and Whipsnade Wild Animal Park, but the Federation of Zoological Gardens of Great Britain and Ireland publishes a complete list of its 60 members. In addition there are another 200 wild animal collections in Britain.

Conditions of work vary and there is no set career development path or recruitment strategy. Likewise formal educational requirements differ from place to place and it is quite possible to get work without any. Most employers insist on a probationary period before they support an individual's training programme and it is normally advantageous for applicants to have shown some commitment to animals, either through voluntary, seasonal or weekend work.

Qualifications and training

No academic qualifications are normally required, though a background of natural science at school, and an understanding of animals, are an advantage, as are relevant A levels and/or a university degree. Recruits to the keeper staff at London Zoo or Whipsnade Park are expected to have some previous experience of working with animals and a good general education. But the competition is so fierce they usually only recruit people who have some proven ability to work with animals. This could be in a pet shop, vet surgery or another zoo. The Zoological Society's keeper staff take the City and Guilds course in Zoo Animal Management which is run by the National Extension College. This is a correspondence course supplemented by the Society with additional on-site lectures and trips to other zoos. Training arrangements, as well as salaries, vary between different zoos. Vacancies for London Zoo are usually advertised in specialist magazines, such as *Cage and Aviary Birds*. Alternatively, you can apply in writing to the Senior Personnel Officer.

What the job is really like

Contact with animals is only a small part of the day's work, which is much more taken up with cleaning out and preparing endless amounts of food. Beautiful friendships with the animals are rare. The creatures are wild and usually dangerous, the hours unsocial and the pay is not marvellous. So why do so many thousands of youngsters find this job attractive?

Senior keeper

'For a kid who really knows what he or she is taking on, who is prepared for the hard slog that each day brings, and who has no romantic ideas of a glamorous job', said a senior keeper, 'it's a good career. But I've talked to a lot of them who think they'd like the work and most of them had a completely unrealistic picture. They've seen the camel rides or elephants' bath-time and think keepers spend their time playing games with the animals. Or else they have what a colleague of mine calls "Elsa fantasies", seeing themselves striking up a sympathetic relationship with a big cat. Believe me, the only relationship you get with a lion or a tiger is mutual respect.'

Affection for animals

An affection for animals is essential but, by itself, not enough. The same keeper said: 'There's no point in aiming for zoo work unless you have a strong feeling for animals, but that's not always what people mean when they say they like them. You need more of an interest in them than a sentimental approach.

'Now, I haven't got a dog, but if I had one, or a cat, or any domestic pet I would need a completely different attitude to it from the one I would have towards any animal in this zoo. I'd train it to behave in a manner acceptable to humans, for instance. You just can't take the same approach to wild animals, even when they appear quite tame. And you shouldn't be looking for affection from them. Respect now, that's another matter. You must have that and you have to earn it, and for me there's more reward in that.'

Routine

The day starts at 8 am and can go on as late as 7 pm. A zoo will insist on punctuality, not least because the animals do. They become accustomed to routine and looking after them is a disciplined business. It is also extremely hard work.

Scraping bird cages on a frosty morning or shovelling manure and heaving huge bundles of straw is not the side of zoo work which usually occurs to someone applying for a job. Even someone who should know can be taken aback by it. Daryl is 17, a zoo helper, and he should have had few illusions as his father works at the Zoo. But he admits: 'I'd no idea what mucking out a cage was really like.' And Daryl works with small mammals. Think of the lions' cage first thing in the morning.

It is not a job for the physically frail, male or female, for it has to be done quickly and thoroughly.

Case Study

Ann is 27 and a qualified keeper, working in the Bird House.

'A girl has to be prepared to work harder than the men to prove herself. I know I do. It's been a man's domain for so long. I suppose it's understandable they take a rather superior attitude to women. But a strong girl can work just as well as a strong chap. Oddly enough, although it sounds daft, I agree animals can tell the difference between the sexes. I worked in a parrot house where some of them preferred men and tried to attack me. But then they were used to men. The other thing animals recognise is authority, and if you've got that they'll respect you, male or female.

You've got to have a real vocation for this work, but you can't be sentimental. I always wanted to work with animals. My first job was with a bird farm in Keston. I've worked in RSPCA clinics and when I was an assistant I was sent on a training scheme, helping a vet for six months. I quite enjoy this work although I wouldn't mind a spell with small animals. The funny thing is, the talking birds do actually say 'Hello'. I suppose it's because that's what everyone says to them.

But really I enjoy the job. I've been here for three years and I think things are improving. There are more chances now for girls than there were.'

Typical day

A typical day in a keeper's life moves on from cleaning to food preparation. Again, this can be physically demanding in some cases, such as in the Elephant House. Each elephant needs a hundredweight of hay every day as well as vegetables and pellet food. If an animal is refusing food a keeper has to spot that quickly and find out why. Cleaning and chopping food can occupy a surprisingly large part of a working day, particularly in the Bird House where one tray can call for more than a dozen different sorts of food. Often supervision is needed at feeding time, particularly if there is a specially dominant animal which would take more than its share.

Some creatures require hand feeding. The king penguin will not eat dead fish unless it is placed in its mouth. Some animals show off to attract a keeper's attention. For others, meal-times are exercise sessions. This is one of the reasons a keeper throws fish to sea-lions, keeping them on the go and making sure each gets a fair share.

Visitors

Keepers must always be aware of visitor activity. Before animals were completely isolated from visitors, keepers kept a record of the number of items 'lifted' from unsuspecting visitors. In one year alone, the elephants seized 14 coats, 12 handbags, ten cameras, eight gloves and six return tickets to Leicester, damaging them beyond repair.

Now the elephants are kept beyond arm's reach, the keepers

have fewer irate complainers to deal with. Even so, with over a million visitors a year they must be able to look after the human species as well as they do the others. As well as taking part in public presentations on animal matters, keepers must be ready to answer questions on more general topics, always answering politely and intelligently.

Further qualifications

The National Extension College course in Zoo Animal Management is intended for keepers responsible for the care of animals in a zoo or wildlife collection, either in this country or abroad. It is open to any student of 16 or over. The course combines correspondence tutoring with practical work, directed by a tutor at the student's own zoo and periodic practical assessments. Those who do not have a job in a zoo or wildlife collection can still take the correspondence part of the course but will not be able to enter for the examination for the City and Guilds Certificate in Zoo Animal Management.

The course lasts two years and covers various aspects of zoo management including children's zoos, finance, housing, zoo design, nutrition and diet, breeding and animal behaviour. Specific animal groups are studied, from fish and reptiles to birds, primates, elephants, carnivores, and the vital topic of endangered species.

Salaries

Wages vary according to age and experience and between different establishments. London Zoo offers an incremental pay scheme which rises according to experience. Trainee keepers receive £10,895 a year for three years, in which time they should complete the City and Guilds Certificate in Zoo Animal Management. On completion of the training period, they become qualified keepers and earn £13,925 a year. Qualified keepers who have at least three years' experience can apply to become senior keepers with a salary of £15,824.

Making a start

Not all members of the Federation of Zoological Gardens of Great Britain and Ireland are zoos; for example, Knowsley Safari Park, Liverpool Museum Aquarium and Vivarium, the Wildfowl and Wetlands Trust and the Dartmoor Wildlife Park.

Recruitment varies from place to place and potential employers need to be contacted individually. Enquiries should always be accompanied by a stamped addressed envelope. For advertised vacancies, look in *RATEL*, the journal of the Association of British Wild Animal Keepers (ABWAK), the weekly paper *Cage and Aviary Birds, International Zoo News* and local newspapers and JobCentres.

As an example London Zoo employs seasonal staff every summer. The best of these may be taken on to become keepers, but as the Zoo has thousands of applicants every year, competition is stiff. If you can't get a seasonal keeper's job perhaps you could consider spending some time working as a general attendant, or as a volunteer. Contact your nearest zoo for details. The Federation will provide you with a list.

It is possible to work overseas, but usually only with good references and a lot of experience.

9 Agriculture

Agriculture in Britain has come under great scrutiny in recent years. Some farming practices have had to change in the light of new discoveries, most notably BSE, and many people have started to question the sort of modern agricultural techniques that have been held responsible. Anyone becoming a farmer not only has to contend with the day-to-day farming experience, which is arduous and exhausting, but must also be capable of dealing with the myriad problems associated with modern farming, which include fluctuating prices and market conditions, livestock subsidies, European legislation and the growing concerns of the sceptical consumer. Not only this, some farmers have taken the decision to switch to organic production, with its lower yields and greater costs. As interest in organic food increases, this may well be a decision which many farmers will face in the coming years.

Not that this puts off the many new entrants to farming each year. This is perhaps because agriculture is a way of life, with its own community. For some, farming is in the blood, but there are also newcomers, and although the chances of owning a 1,000 acre farm are practically nil for someone working from a standing start, it is not impossible to enter farming from the outside.

Some farmers have a variety of animals but most farm only one species. An agricultural worker will generally specialise and remain a herdsmen or a shepherd or a pig farmer throughout their working life, and quite often beyond. As with any craft worker it is with some regret that a farmer gives up putting to use knowledge learnt over decades.

Agriculture

Case Study

Iain was an assistant shepherd in Scotland for three years, and is now seeking a job as a shepherd in Yorkshire.

'I had a very good teacher, a shepherd who had been working with sheep for 42 years. He didn't have a very high opinion of their IQ, although he always said they weren't as stupid as they looked, and he basically liked them.

He worked out a method, during the breeding season, of checking that all the ewes had been served by the rams and I told him he should patent it. In fact, you have to have a method of marking the ewes to make sure they will produce lambs. There's always a dominant ram and, if you don't watch, there are some rare fights. You can't allow that, of course. They can damage each other and they are valuable beasts. I've been asked whether I'm afraid of rams. The answer is, I'm careful and I have a good dog. To be truthful I would as soon cope with a ram as a sheep in a bad temper.

The real excitement, of course, is the lambing. You don't get that much sleep during the lambing and losing a ewe is a calamity, even to experienced shepherds. Most people know how difficult it can be then, to persuade another ewe to rear the lamb of a dead mother. There were two wee girls on the farm where I worked and every season they'd hand-rear an orphan or two. I sometimes used to think they would pray for a lamb to be bottle-fed and there were the shepherds saying their prayers that there wouldn't be any orphans.

During the winter we take crops out to the flock, and big blocks of salt. I remember one bad year when we had to hand-feed them. Not all the ewes are suitable for breeding, and you have to learn how to select the stock.

It's a grand feeling when the lambing is over – if it's been a good season. But you don't put on your hat and take off for the Bahamas. They need checking all the time. There's a lot can go wrong with sheep and yes, they do roll on to their backs and get stuck. It's the weight of the fleece. You'd be surprised the people who think you just herd a flock out to the hills and leave them there.

When I started, I thought I'd give it a year. Now I'm really hooked on the whole thing.'

Farm sizes vary enormously but however many animals you tend, the round of jobs is always the same. Take the example of a dairy herdsmen.

Careers working with animals

Case Study

James works on a West Country farm with a herd of 60 cattle, small by modern standards but quite enough for him.

'There's an old song that sums it up: 'I've got the milk 'em in the morning, feed 'em, milk 'em in the evening, blues.' There's a lot of job satisfaction in looking after a herd, but it certainly is demanding. It means getting up about 4 am and you can't just go back to bed if it's raining or you had a heavy night the night before. If you've ever heard cows complaining about a late invitation to the milking parlour, you'd know what I mean.

I use a dog to help herd them in, in summer. She's a cross-bred collie and the cows respect her, but I have one who will stand up to any dog and makes a point of taking her own way in her own time. She does it to keep the respect of the herd. They have a very complex social structure, cows, and you learn to respect it.

In winter they are indoors so it's easier in the mornings and the milking parlour can seem quite cosy on a cold rainy morning, with the radio tinkling away and the milk sloshing beautifully – that is, when nothing's gone wrong with the equipment. It's a large part of my job to make sure nothing does. Each cow has to have its udders washed before the milking cups are attached. They're all used to the routine, although I've heard some pretty hilarious stories from old hands about the introduction of milking machines.

My wife gives me a hand usually. You've got to remove the milking cups pretty promptly when the animal has given all she intends to, so you have to keep a close watch on the jars.

Our milk goes into a tank, and that has to be 100 per cent clean. So does the equipment, which is sterilised after each use, and the dairy has to be swabbed out each time too ... and all this before breakfast. Well, some dairymen might have time for breakfast at 4.30 but not this one, and even if I did, I'd be ready for another after milking.

It's not all over after the morning session. I do a lot of mucky work and a lot of heavy fodder and bedding around. We rear our own calves here and I'm usually there at a calving. But it's not tiring the way a city job can be, and I know because I've done both. You don't get drained doing this work. Exhausted, yes, but a healthy exhaustion. I like the work and I like the animals. They're characters, cows.'

For those wishing to find out more, the Dairy Council produce a package called 'Looking Ahead – Careers in the Dairy Industry'. It should be available from your careers officer or direct from the National Dairy Council.

Those starting their careers will need experience first. Sometimes this comes down to visiting your local farmer, or arranging through your school a period of work experience on a farm. Maintaining contact with your local NFU secretary, agricultural markets and the local Young Farmers Club is also a good idea. If you live in a city it is still possible to get some experience on a city farm, although the majority of farm workers come from rural areas. ATB Landbase advise those living in urban areas to contact their nearest College of Agriculture/Horticulture and consult the *Farmers Weekly* and *Farmers Guardian* (North and Wales). Those interested in organic techniques can find useful experience through an organisation called **Willing Workers On Organic Farms (WWOOF)**. Find out if there are any organic farms in the area.

Not only will such initial work experience help you make up your mind whether or not farming is the right career for you, it will help to prepare you for proper training.

Training

Local branches of the Agricultural Training Board (ATB), agricultural colleges, Training and Enterprise Councils (TECs) and Jobcentres can all give advice on training opportunities.

Training is given through full-time and day- or block-release college courses, and Youth Training. Trainees can study for qualifications in agriculture, livestock or farm management, up to degree level. Agricultural colleges have large farms attached and also offer short, part-time and evening courses in specialist subjects. Full-time working experience in two or more farming enterprises is an entry requirement for many college courses; this can start during school holidays and weekends with unpaid work on a local farm.

The agreement of the local education authority is needed

before enrolment in a college outside the area in which you live; grant aid is not normally available for part-time courses.

No specific GCSE passes are needed for entry to BTEC* First Diploma and NEB First Certificate courses in Agriculture, but an acceptable standard in English and maths will be expected. Candidates for National Certificate courses should have at least one (and preferably two) years' relevant practical experience, or have completed a First Diploma course. After further experience, trainees can take the Advanced National Certificate (ANC). The National Certificate in Farm Management requires a minimum age of 20 years on 1 September of the year of entry, a previous National Certificate (NC) in Agriculture course (or its equivalent) and at least two years' practical experience (of which one year should follow the NC course).

Normal entry requirements for BTEC National Diploma courses are: minimum of four GCSE passes and one year's experience; BTEC First Diploma; NEB National Certificate; relevant qualifications gained through part-time study and one year of practical experience.

GCSE and A-level passes are needed for entry to a Higher National Diploma course. Entry to a degree course requires at least two A-level science subjects and good GCSE subject passes comparable with other science-based disciplines.

A list of full-time and sandwich courses available in the UK is given in the *Directory of Courses in Land Based Industries*, £19.95 from Farming Press Books, Wharfedale Road, Ipswich IP1 4LG; 01473 241122. For further information about careers in farming you can write (enclosing large sae) to ATB Landbase, or phone them on 0345 078007, charged at local rate.

As farms increase in size there is a trend towards specialisation, so it is a good idea to aim to become a skilled stockman, shepherd, pigman, poultryman or cowman. Skilled craftsmen command high wages and can become supervisors or unit managers and even, eventually, farm managers. Taking advantage of any courses throughout your career will be very helpful.

Please note:
  * BTEC is governed by the **Edexcel** foundation, and all enquiries should be directed to the office in London.

Modern Apprenticeships

ATB Landbase have come up with a structured programme for training on the job within the framework of a Modern Apprenticeship (a full description of which can be found in Chapter 12). Three sets of objectives are set at the start: Foundation, Intermediate and Final. By the end of the Final target apprentices will be sufficiently proficient to qualify for an NVQ level 3 award in, for example, Livestock Production. Full details of the arrangement are available from ATB Landbase in their factsheet, *Modern Apprenticeships in Agriculture*.

National Vocational Qualifications (NVQs)/Scottish Vocational Qualifications (SVQs)

Other employees will prefer to offer training without the Modern Apprenticeship framework. Trainees will still be encouraged to take one of the NVQs available:

Agriculture (Crop and Livestock Production) levels 1, 2 and 3
Agriculture (Hatchery Production) level 2
Agriculture (Livestock Management) level 4
Agriculture (Livestock Production) levels 1, 2 and 3
Fish Husbandry level 3
Fish Husbandry (Fin Fish) level 2
Fish Husbandry (Shellfish) level 2

Pay

The basic wage for Grade 1 skilled workers in the industry as laid down by the Agricultural Wages Board is £209.32 per week (1996). Workers may receive more than this, but it relates to their responsibilities and the extra hours they work; overtime is £8.05 per hour for a Grade 1 worker. Up-to-date information on wages can be obtained from the Agricultural Wages Board. This is one of the top levels of pay, and wages differ depending on skills and age. A sixteen-year-old, for example, can only expect to earn about £95.00 a week.

Fish farming

According to ATB Landbase 5,000 people are now employed in fish farms. It is one of Britain's growing industries, with rainbow trout and Atlantic salmon, oysters, mussels and scallops being the most common crop. Although qualifications may not be essential, they are useful and further details can be obtained from the Institute of Fisheries Management.

Organic farming

Organic farms live and work by a different set of rules and practices. As a result organic farmers need different training and experience. For those who are new to the field it can appear confusing, but there are specific training programmes and lots of opportunity to gain first-hand experience.

Like all other forms of agriculture, organic farming is a lifestyle as much as a job. Successful enterprises like Racheal's Dairy in Wales have managed to carve a niche in the market, offering high quality produce at prices above the norm. Often marketing is the key to success for many organic farmers and a certain amount of business acumen is essential if you are to succeed. Some people are able to complement their income from one job with an organic smallholding. This allows incomes to be more flexible and provides individuals with an opportunity to farm animals for their own private consumption.

There are a number of organisations offering advice and information about organic farming but you could start off by contacting the Organic Food and Farming Centre at the Soil Association, for a list of courses.

City farms

Animals have always made good tourist attractions and many farms have opened their gates to the public to earn more money, or to provide educational resources for local school children. City

Agriculture

farms in particular have given many children their first glimpse of pigs, chickens and horses. They provide a valuable resource for cities and although the potential for full-time employment is small there is always work for volunteers. Contact the National Federation of City Farms for details.

10 Getting in and getting on

Top Tips

for Getting into Animal Work

- Few employers are prepared to take on someone who has no experience, training or proven interest in their chosen career. By proven interest I mean tangible proof that you have worked with animals in a way that demonstrates you have an affinity with the animals you care for. This could be as straightforward as looking after pets in your spare time.
- Enterprise is always appreciated. Case studies in this book included the pet shop owner who turned a holiday job into a career and ended up becoming the manager of a chain of stores and the dog trainer who got a job by demonstrating to her employer that she could train a pup.
- If you don't already have experience, and many people interested in a career working with animals do, you should look at the options open to you. First of all you will need to make a choice. Out of all the careers mentioned in this book which one are you most suited to, considering your personality, skills and attributes? Deciding what sort of animal work you want means finding out quite a bit about yourself.

- If you haven't already considered volunteering, do so. Not only is this a good way of finding out more about yourself, it often provides the first contact with a potential employer and gives you an opportunity to learn about the job you are hoping to do.
- Potential employers will be impressed by individuals who give up their time for free, particularly if you have had to suffer some material hardship, putting off the chance to earn money for example. In some cases, this is the only way of making an impression.
- Some people may ask you to prove yourself by working for them for nothing first. Take care not to be exploited. Any commitment on your part should be for short periods only. A week or two at the most. Some organisations run legitimate schemes which may last up to six months. The RSPB and London Zoo are good examples of these, where a period of voluntary work is the only way to make your abilities known to the employer.
- Not all jobs are like this, though, and you should always check out the specific demands of each career. The best people to talk to are the organisations listed in this book. Find out from them what is expected of you. They will usually have an appropriate information sheet/pack and may be able to spend some time talking to you.
- They may also be able to give you valuable contacts. As in all employment sectors, contacts are as important as qualifications, experience and ability. Many jobs aren't advertised and those who are lucky have often made their own luck, finding themselves 'in the right place at the right time'. Sometimes a chance letter to a potential employer can do more good than a dozen applications.

11 The future of animal work

A career working with animals will continue to be a popular choice among school leavers. In substance the work is unlikely to change dramatically. Animals will always require care and attention. More likely to change are the techniques with which we offer this care and attention.

Veterinary surgeons will continue to update the type of treatment they offer sick animals, whether this be through conventionally applied medicine or in the new field of animal homeopathy (the treatment of an animal as a whole rather than the treatment of a specific complaint). The profession maintains a balanced entry from year to year and this is unlikely to change in the foreseeable future. The same can be said for veterinary nurses, whose employment is directly related to the number of vets.

The number of animal welfare organisations has increased as a response on the one hand to a greater level of consciousness regarding the humane treatment of animals, and on the other hand a seemingly continuous rise in the number of reported cases of animal abuse. Incongruously, at the same time we seem to care more and less about our animals.

Animal conservation has become a topic of much debate, as the human desire for material wealth continues to conflict with the survival of animals in the wild: not just in Britain where wildlife is threatened by a sustained building and roads programme, but abroad in areas where the policies of developing countries threaten endangered species still further. Even so, if you compare this sector with some of the others in this book, such as agriculture, opportunities are much rarer and harder to come by. The

The future of animal work

sector lacks the training structures associated with some of the other sectors and entry quite often happens in a much more unpredictable way.

In a time of great technological advance it is reassuring to know that the number of individuals working with dogs and horses continues to grow. The two sectors are likely to improve their training structures, incorporating more National Vocational Qualifications and Modern Apprenticeships. The National Horse Education and Training Company hope to build up their training programme to include new units in the existing Horse Care and Management NVQ for side saddle, polo craft and assisting the disabled rider. The last component reflects an increased interest in riding therapy. Although the National Lottery is said to have reduced gambling on horses, no immediate job losses are expected and interest in the sport remains strong. There has also been a sustained growth in interest in greyhound racing.

The number and size of zoos continue to shrink, as rising costs combine with lower visitor figures, with a consequential loss of jobs. Following a number of job losses at London Zoo, Britain's biggest employer, the competition for places has become even more intense than it already was. Zoos are beginning to re-evaluate their role as they attempt to establish themselves as a key player in the battle to conserve endangered species through in-captivity breeding programmes. An estimated 20 per cent of all zoos are currently engaged in what the Federation of Zoological Gardens describe as serious research to breed species in captivity.

The pet trade continues to expand; check out the new range of services offered by some entrepreneurs, including pet astrology and pet funerals and the new pet supermarkets. The Japanese have invented the computerised pet. It is a small computer you keep in your pocket. You must remember to keep it well fed, clean and happy. Just like a real pet, if you forget about it, it dies. Unlike a real pet, you can just turn the computer back on again and start the process all over. The Pet Care Trust, which acts as an industry watch dog and training provider, does not predict any significant changes in the next two or three years.

Agriculture in Britain has faced a number of crises in recent years, not least of all BSE. The beef industry say they have been set back years and some farmers, forced to feed cattle waiting for the cull, have faced severe financial hardship. Whether these problems will continue to grow is very much a matter of debate, with one side proclaiming business as usual and the other demanding substantial change. One thing is for sure: the interest in organic beef and dairy products has risen so sharply that a large proportion of organic produce has to come from abroad, a situation that you would expect to be rectified over the next few years. The number of people working in agriculture and horticulture remains at about 600,000 (2 per cent of the population) and although the industry believes there are always more interested applicants than places, as an employment sector it is less competitive than others in this book.

12 Qualifications available

Many jobs require at least some level of general academic qualification before applicants can start to undertake vocational training. Although it is still possible to get work without them, the accepted entry point standard is four GCSE passes at C grade or above, two of which should be maths and English.

This is not always the case and this book has already listed examples of courses which allow you to reach the equivalent academic standard. Pre-vocational courses enable those who do not meet the GCSE standard, but nevertheless show a vocation for the job, to start vocational training, which will lead to full-time employment. For example, veterinary nurses can take a pre-vocational qualification at 16 and vocational training at 17, without losing any time over GCSE students.

Vocational training usually starts at 16 and A-level qualifications aren't required for most jobs. One of the few exceptions to this is the vet, who will have very good A-level results before beginning the long degree course. Some areas of animal research also demand high academic standards and agricultural students can take a vocational route through university.

For most people, training begins on the job between the ages of 16 and 18. Most of the training is free and in some regions students are given credits to buy the training they want. These credit schemes vary depending on where you live, so contact your local careers office who will have full details of schemes in your area. There are a number of different training options to choose from.

National Vocational Qualifications (NVQs) and Scottish Vocational Qualifications (SVQs) are the new post-16 industry training standard for school-leavers. They are assessed in a work situation, either at a college or in a job. Students take study units, some based on practical skills and others based on written work, with a range of requirements.

Different units can be built up into a complete NVQ/SVQ. Most colleges offer NVQ/SVQ courses, and the qualifications are designed to give employers an idea of a job candidate's capabilities, both in the UK and, eventually, throughout Europe. There are five levels of qualification – one to five – and each individual sector will have its own expected standards.

The following NVQs are available in England and Wales: levels 1, 2 and 3 Agriculture (Crop and Livestock Production); level 2 Agriculture (Hatchery Production); level 4 Agriculture (Livestock Management); levels 1, 2 and 3 Agriculture (Livestock Production); level 1 Animal Care; level 3 Animal Training; level 3 Animal Welfare and Management; level 2 Caring for Animals; level 3 Fish Husbandry; level 2 Fish Husbandry (Fin Fish); level 2 Fish Husbandry (Shellfish); levels 2 and 3 Gamekeeping; levels 1 and 2 Horse Care; level 3 Horse Care and Management; level 3 Pet Care and Supply; level 2 Racehorse Care; level 3 Racehorse Care and Management.

The following SVQs are available in Scotland: level 1 Animal Care; level 3 Animal Training; level 3 Animal Welfare and Management; level 2 Caring for Animals; levels 1, 2 and 3 Crop and Livestock Production; levels 2, 3 and 4 Environmental Conservation: Landscapes and Ecosystems; levels 2 and 3 Fish Husbandry; levels 2 and 3 Gamekeeping; levels 1 and 2 Horse Care; level 3 Horse Care and Management, level 4 Livestock Management; levels 1, 2 and 3 Livestock Production; level 2 Racehorse Care; level 3 Racehorse Care and Management; level 3 Pet Care and Supply.

General National Vocational Qualifications are designed to offer students a broadly based qualification which is something of a cross between academic and vocational training programmes. The courses are normally full time and offered at three levels: Foundation, Intermediate and Advanced. Whereas the NVQ will prepare you for a specific job, the GNVQ allows you to keep your options open, preparing you instead for a range of jobs. The GNVQ Land and Environment is a modular course with a number of relevant modules related to caring for animals. Contact the GNVQ helpline for details: 0171 728 1914.

Neither of these has replaced the BTEC Certificate and Diploma or Scottish Certificate and Diploma courses, which offer a similar training pattern, as students advance from National Certificate to the Higher National Diploma and degree. Agricultural students often choose this route as it combines college taught management skills with practical farm experience. Most agricultural colleges have their own farms so it is still possible to get good 'hands-on' experience. For a full list of course units available, students should refer to the publications produced by BTEC and the Scottish Qualification Authority (formerly SCOTVEC).

National Certificate (NC) courses last for one year and anyone over 17, with at least one years practical farming experience, is eligible. Holders of the NC, or those with four GCSE passes (grades A–C), are then eligible to gain entry to a National Diploma (ND) course. These are organised on a two-year full-time or three-year sandwich basis, with the middle year spent working. The next stage up, the BTEC Higher National Diploma is science-based, designed as it is for students who wish to work in posts demanding managerial and technological skills. You must be aged 18 or over and have either one A-level in a science subject, a BTEC National award or a level 3 NVQ or GNVQ. The course lasts two years full time or three years sandwich.

Modern Apprenticeships are designed for 16- and 17-year-old school-leavers. They offer the apprentice training to at least NVQ level 3, which is roughly equivalent to two A-levels or an Advanced GNVQ, and a small wage. Each apprenticeship lasts for about three years and the apprentice follows a training plan which is agreed at the start. By the end the apprentice will have learned the skills needed to work in the chosen industry, and training also includes skills such as team work, communication and problem solving.

The horse industry has taken the lead and offer Modern Apprenticeships through the National Horse Education and Training Company, from which a number of useful leaflets are available. ATB Landbase has organised a Modern Apprenticeship programme for the agricultural sector. Ask for their factsheet number 22 (22a if you live in Scotland). Modern Apprenticeships are usually organised on a local basis and your local Training and Enterprise Council (TEC) or Local Enterprise Company (LEC) in Scotland should already be setting up schemes in your area. If there isn't already an established contact for animal work, encourage them to look for one by showing that you are enthusiastic. Enthusiasm always impresses people. You might even be able to persuade a local employer yourself that you are a suitable candidate.

Accelerated Modern Apprenticeships are designed for 18- and 19-year-olds who may wish to build on skills and knowledge they already have. Again, there is a wage but this time the training programme takes between 18 months and two years only.

13 Where to study

This chapter contains a list of colleges. Use it to find out where you can get qualified. The first section is for veterinary surgeons only. Section two lists colleges which provide courses in Animal Care, Veterinary Nursing, Agriculture and Equine Studies. The third section is for those considering a correspondence course.

Section one: degree courses in veterinary science

Bristol
Veterinary Admissions Clerk, University of Bristol, Senate House, Bristol BS8 1TH; 0117 928 7679

Cambridge
The Department Secretary, Department of Clinical Veterinary Medicine, University of Cambridge, Madingley Road, Cambridge CB3 0ES; 01223 337600

Edinburgh
The Dean, Faculty of Veterinary Medicine, Royal (Dick) School of Veterinary Studies, University of Edinburgh, Summerhall, Edinburgh EH9 1QH; 0131 650 6130

Glasgow
The Chairman of the Admissions Centre, University of Glasgow, Veterinary School, 464 Bearsden Road, Bearsden, Glasgow G61 1QH; 0141 330 5705

Liverpool
The Admissions Tutor, Faculty of Veterinary Science, University of Liverpool, PO Box 147, Liverpool L69 3BX; 0151 794 4281

Careers working with animals

London
The Registrar, The Royal Veterinary College, University of London, Royal College Street, London, NW1 0TU; 0171 387 2898

Section two: college courses

A number of colleges offer courses in Veterinary Nursing, Animal Care, Equine Studies and Agriculture. The following is a list of these colleges. Each college is listed under county names so you can easily find out which one is closest to you. After the college name is a list of acronyms. These indicate the type of courses available. The following key is used:

Agriculture - **Ag**   Stud work - **Eq(s)**
Animal Care - **AC**   Veterinary Nursing - **VN**
Equine Studies - **Eq**

The name Equine Studies is a catch-all and colleges will provide exact details of courses on offer, but they may include the British Horse Society examinations, BTEC and SCOTVEC Certificates and Diplomas, NVQs and SVQs, GNVQs and a number of other related qualifications. Unfortunately, space does not permit a complete course listing.

College listing by county

Berkshire
Berkshire College of Agriculture **VN, AC, Eq, Ag**
Hall Place, Burchetts Green, Maidenhead SL6 6QR; 01628 824444

Birmingham
Matthew Boulton Technical College **VN**
Sherlock Street, Birmingham B5 7DB; 0121 446 4545

Bristol
University of Bristol School of Veterinary Science **VN**
Langford BS18 7DU; 0117 928 9286

Cambridgeshire
Cambridgeshire College of Agriculture and Horticulture **VN, AC, Eq(s), Ag**
Landbeach Road, Milton, Cambridge CB4 6DB; 01223 860701

College of Animal Welfare **VN, AC**
Wood Green Animal Shelters, Godmanchester, Huntingdon PE18 8LJ; 01480 831177

Peterborough Regional College **Eq**

Where to study

Park Crescent, Peterborough PE1 4DZ; 01733 767366

Cheshire
Reaseheath College **AC, Eq**
Reaseheath, Nantwich CW5 6DF; 01270 625131

Cleveland
Stockton/Billingham College of Further Education **Eq**
The Causeway, Billingham TS23 2DB; 01642 360205

Cornwall
Duchy College of Agriculture **AC, Eq, Ag**
Stoke Climsland, Callington PL17 8PB; 01209 843722

Coventry
Coventry Technical College **VN**
The Butts, Coventry CV1 3GD; 01203 526700

Cumbria
Newton Rigg College **AC, Eq, Ag**
Newton Rigg, Penrith CA11 0AH; 01768 863791

Derbyshire
Broomfield College **VN, Ag**
Morley, Ilkeston, Derby DE7 6DN; 01332 831345

Devon
Bicton College of Agriculture **VN, AC, Eq, Ag**
East Budleigh, Budleigh Salterton EX9 7BY; 01395 568353

East Devon College **Eq, Ag**
Bolham Road, Tiverton EX16 6SH; 01884 254247

Dorset
Bournemouth & Poole College of Further Education **VN**
North Road, Parkstone, Poole BH14 0LS; 01202 747600

Kingston Maurward College **VN, AC, Eq, Ag**
Dorchester DT2 8PY; 01305 264738

Durham
Houghall College of Agriculture and Horticulture **Eq, Ag**
Houghall, Durham DH1 3SG; 0191 386 1351

Essex
Writtle College **Eq, Ag**
Writtle, Chelmsford CM1 3RR; 01245 420705

Gloucestershire
Hartpury College **AC, Eq, Ag**
Hartpury House, Hartpury GL19 3BE; 01452 700283

Royal Agricultural College **Eq, Ag**
Cirencester GL7 6JS; 01285 652531

Hampshire
Farnborough College of Technology **VN**
Boundary Road, Farnborough GU14 6SB; 01252 515511

Sparsholt College **AC, Eq, Ag**
Sparsholt, Winchester SO21 2NF; 01962 797280/776441

Hereford
Holme Lacy College **AC, Eq, Ag**
Holme Lacy HR2 6LL; 01423 870316

Hertfordshire
Oaklands College **VN**
The Campus, Welwyn Garden City AL8 6AH; 01707 326318

Oaklands College **AC, Eq, Ag**
Oakland Campus, Hatfield Road, St Albans AL4 0JA; 01727 850651

Kent
Hadlow College of Agriculture and Horticulture **Eq, Ag**
Hadlow, Tonbridge TN11 0AL; 01732 850551

Springboard Bromley **VN, AC, Eq**
Bromley College, Rookery Lane, Bromley BR2 8HE; 0181 462 1222

Thanet College **Eq**
Ramsgate Road, Broadstairs CT10 1PN; 01843 860482

Lancashire
Myerscough College **AC, Eq, Ag**
Myerscough Hall, Bilsbarrow, Preston PR3 0RY; 01995 640611

Leicestershire
Brooksby College **Eq, Ag**
Brooksby, Melton Mowbray LE14 2LJ; 01664 434291

Lincolnshire
De Montfort University Lincoln **AC, Eq, Ag**
School of Agriculture & Horticulture, Caythorpe Court, Caythorpe, Grantham NG32 3EP; 01400 272521

North Lincolnshire College **Eq**
Monks Road, Lincoln LN2 5HQ; 01522 510530

London
Kingsway **VN**
Sidmouth Street, Gray's Inn Road, London WC1H 8JB; 0171 278 0541

Cordwainers College **Eq (saddlery)**
182 Mare Street, Hackney, London E8 3RE; 0181 985 0273

Manchester
North Trafford College of Further Education **VN**
Talbot Road, Stretford, Manchester M32 0XH; 0161 872 3731

Merseyside
City of Liverpool Community College **VN**
Muirhead Avenue East, Liverpool L11 1ES; 0151 252 1515
Mabel Fletcher Centre, Sandown Road, Liverpool L15 4JB; 0151 252 1515

Middlesex
Capel Manor **AC, Eq, Ag**
Horticultural and Environmental Centre, Bullsmoor Lane, Enfield EN1 4RQ; 0181 366 4442

Norfolk
Easton College **AC, Eq, Ag**
Equestrian Centre, Easton, Norwich NR9 5DX; 01603 742105

Norfolk College of Arts and Technology **Eq**
Tennyson Avenue, King's Lynn PE30 2QW; 01553 761144

Northamptonshire
Moulton College of Agriculture and Horticulture **Eq, Ag**
West Street, Moulton, Northampton NN3 7RR; 01604 491131

Tresham College **Eq Ag**
Kettering Centre, St Mary's Road, Kettering NN15 7BS; 01536 410252

Northumberland
Kirkley Hall College **VN**
Ponteland NE20 0AQ; 01661 860808

Nottinghamshire
Brackenhurst College **AC, Eq**
Brackenhurst, Southwell NG25 0QF; 01636 817000

Oxfordshire
West Oxfordshire College **Eq(s), Ag**
Holloway Road, Witney OX8 7EE; 01993 703464

Shropshire
Walford College **Eq, Ag**
Baschurch, Shrewsbury SY4 2HL; 01939 260461

Somerset
Cannington College **Eq**
Cannington, Bridgewater TA5 2LS; 01278 652226

Staffordshire
Rodbaston College of Agriculture **VN, AC, Eq, Ag**
Penkridge, Stafford ST19 5PH; 01785 712209

Careers working with animals

Suffolk
Otley College **Eq, Ag**
Otley, Ipswich IP6 9EY; 01473 785543

Surrey
Merrist Wood **AC, Eq, Ag**
Worplesdon, Guildford GU3 3PE; 01483 232424

NESCOT (North East Surrey College of Technology) **VN**
Reigate Road, Ewell LT17 3DS; 0181 394 1731

Sussex
Brinsbury College of Agriculture and Horticulture **VN, Ag**
North Heath, Pulborough, West Sussex RH20 1DL; 01798 873832

Plumpton College **Eq, Ag**
Plumpton, Lewes, East Sussex BN7 3AE; 01273 890454

Warwickshire
Warwickshire College of Agriculture **AC, Eq, Ag**
Moreton Hall, Moreton Morrell CV35 9BL; 01926 651367

Wiltshire
Lackham College Joint Equestrian Centre **Eq**
Lacock, Nr Chippenham SN15 2NY; 01249 443111

Worcestershire
North East Worcestershire (NEW) College **VN**
Redditch Campus, Peakman Street, Redditch B98 8DW; 01527 570020

Worcestershire College of Agriculture **Eq(s), Ag**
Hindlip, Worcester WR3 8SS; 01905 451310

Yorkshire
Askham Bryan College **VN, AC, Aq, Ag**
Askham Bryan, York YO2 3PR; 01904 702121

Dewsbury College **Eq**
Halifax Road, Dewsbury, West Yorkshire WF13 2AS; 01924 465916

Bishop Burton College of Agriculture **AC, Eq, Ag**
York Road, Bishop Burton, Beverley, East Yorkshire HU17 8QG; 01964 550481

Rotherham College of Arts and Technology **VN**
Eastwood Lane, Rotherham S65 1EG; 01709 362111

Queen Ethelburga's College (School) Equestrian Centre **Eq**
Thorpe Underwood Hall, Ouseburn, York YO5 9SZ; 01423 330859

Scotland
Borders College **Eq**
Melrose Road, Galashiels TD1 2AF; 01896 757755

Clinterty Agricultural College **Eq, Ag**

Kinnellar, Aberdeen AB5 0TZ; 01224 640366

Edinburgh's Telford College **VN**
Crewe Toll, Edinburgh EH4 2NZ; 0131 332 2491

Oatridge Agricultural College **Eq, Ag**
Ecclesmacham, Broxburn, West Lothian EH2 6NH; 01506 854387

Thurso College **Eq**
Ormlie Road, Thurso, Caithness KW14 7EE; 01847 896161

University of Glasgow **VN**
Veterinary School, 464 Bearsden Road, Bearsden, Glasgow G61 1QH; 0141 330 5700

Wales

Carmarthenshire College of Technology and Art **Eq**
Land-Based Industries Division, Pibwrlwyd, Carmarthenshire SA31 2NH; 01554 759165

Coleg Meiron Dwyfor Glynllifon **Eq, Ag**
Fford Clynnog, Caernarfon, Gwynedd LL54 5DU; 01286 830261

Gwent Tertiary College **Eq**
Usk Campus, The Rhadyr, Usk, Gwent NP5 1XJ; 01291 672311

Pembrokeshire College **AC, Eq**
Haverfordwest, Pembrokeshire SA61 1SZ; 01437 765247

Pencoed College **Eq**
Pencoed, Bridgend, Mid Glamorgan CF35 5LG; 01656 860202

The University of Wales, Aberystwyth/Welsh Institute of Rural Studies **Eq, Ag**
Llanbadarn Fawr, Aberystwyth, Dyfed SY23 3AL; 01970 622021

Yale College **Eq**
Grove Park Road, Wrexham, Clwyd LL12 7AA; 01978 311794

Northern Ireland

Rupert Stanley College of Further Education **VN**
Tower Street, Belfast BT5 4FH; 01232 325312

Enniskillen College of Agriculture **Ag**
Levaghy, Enniskillen, County Fermanagh BT74 4GF; 01365 323101

Greenmount College of Agriculture and Horticulture **Ag**
22 Greenmount Road, Antrim BT41 4PU; 01849 462114

The Queen's University of Belfast, Faculty of Agriculture and Food Science **Ag**
Newforge Lane, Belfast BT9 5PX; 01232 661166

Republic of Ireland

Dublin Institute of Technology **VN**
Kevin Street, Dublin 8

Section three: correspondence courses

Animal Care College
29a Ascot House, High Street, Ascot, Berkshire SL5 7JG; 01344 628269
The Animal Care College is recognised by the Council for the Accreditation of Correspondence Colleges. Students pay a registration fee of £20, which also gives them membership of the Institute of Animal Care Education. No formal qualifications are required before an application can be made except in the case of veterinary nursing, but some courses require candidates to be employed in the industry before they begin. Courses are Veterinary Nursing, Diploma in Pet Bereavement Counselling, Canine/Human Interface, Understanding the Feline/Human Interface, General Certificate of Canine Studies, Judging Diploma, The Dog Breeding Certificate, Grooming for City and Guilds, Kennel Staff and Kennel Management Training, National Small Animal Care Certificate and Diploma of Kennel Management.

National Extension College
18 Brooklands Avenue, Cambridge CB2 2HN; 01223 316644
In conjunction with the National Federation of Zoological Gardens, offers City and Guilds Animal Management (7630).

Pet Care Trust
Bedford Business Centre, 170 Mile Road, Bedford MK42 9UP; 01234 273933
Offers City and Guilds Pet Store Management (7760).

14 Useful addresses

All the following can provide information and advice about a career working with animals. Please send a stamped and self-addressed envelope for replies.

Agricultural Wages Board, Nobel House, 17 Smith Square, London SW1P 3JR; 0171 238 6540

Animal Care College, Ascot House, High Street, Ascot, Berkshire SL5 7JG; 01344 28269

Animal Health Trust, PO Box 5, Shailwell Road, Newmarket, Suffolk CB8 7DW; 01638 661111

The Animal Welfare Trust, Tyler's Way, Watford Bypass, Watford, Hertfordshire WD2 8HQ; 0181 950 8215

Association of British Riding Schools, Queens Chambers, 38–40 Queen Street, Penzance, Cornwall TR18 4BH; 01736 69440

Association of British Wild Animal Keepers, 2A Northcote Road, Clifton, Bristol BS5 6UQ; 0117 973 6480

ATB Landbase, National Agricultural Centre, Kenilworth, Warwickshire CV8 2LG; 01203 696996/0345 078007

Bell Mead Training College for Kennel Staff, Priest Hill House, Priest Hill, Old Windsor, Berkshire SL4 2JN; 01784 431599

The Blue Cross, Shilton Road, Burford, Oxfordshire OX18 4PF; 01993 822651

British Association of Dogs' Homes, Dogs' Home Battersea, Battersea Park Road, London SW8 4AA; 0171 622 3626

British Association of Homeopathic Veterinary Surgeons, Alternative Veterinary Medicine Centre, Chinham House, Stanford-in-the-Vale, Faringdon, Oxon, SN7 8QN

British Dog Groomers Association, Bedford Business Centre, 170 Mile Road, Bedford MK42 9TW; 01234 273933

Careers working with animals

The British Horse Society, Stoneleigh, Kenilworth, Warwickshire CV8 2LR; 01203 696697

British Institute of Professional Dog Trainers, Bowstone Gate, Nr Disley, Cheshire SK12 2AW; 01663 762772

British Racing School, Snailwell Road, Newmarket, Suffolk CB8 7NU; 01638 665103

British Trust for Conservation Volunteers, 36 St Mary's Street, Wallingford, Oxon OX10 0EU; 01491 39766

British Veterinary Nursing Association (BVNA), The Seedbed Centre, Unit D12, Coldharbour Road, Harlow, Essex CM19 5AF; 01279 450567

Cordwainers College, 182 Mare Street, London E8 3RE; 0181 985 0273

The Donkey Sanctuary, Sidmouth, Devon EX10 0NU; 01395 578222

Earthwatch Europe, 57 Woodstock Road, Oxford OX2 6HJ; 01865 311600

Edexcel, Customer Enquiries Unit, Stewart House, 32 Russell Square, London WC1B 5DN; 0171 393 4444

The Environment Council, 21 Elizabeth Street, London SW1W 9RP; 0171 824 8311

Farriers Registration Council and **Farriery Training Service**, PO Box 49, East of England Showground, Peterborough, Cambridgeshire PE2 6GU

Federation of Zoological Gardens of Great Britain and Ireland, London Zoo, Zoological Gardens, Regent's Park, London NW1 4RY; 0171 586 0230/ 0171 722 3333

The Guide Dogs for the Blind Association, Hillfields, Burghfield, Reading, Berkshire RG7 3YG; 01734 835555

Hand to Paw, North Cottage, Great Hayes, Headley Common Road, Headley, Surrey KT18 6NE; 0831 619 847

Institute of Fisheries Management, 22 Rushworth Avenue, West Bridgford, Nottinghamshire NG2 7LF

Hearing Dogs for the Deaf, The Training Centre, London Road (A40), Lewknor, Oxfordshire OX9 5RY; 01844 353898

International Animal Rescue, Animal Tracks, Ash Mill, South Molton, Devon EX36 4QW; 01769 550277

International League for the Protection of Horses, Anne Colvin House, Snetterton, Norwich NR16 2LR; 01953 498682

Useful addresses

International Zoo News Subscription Department, 80 Cleveland Road, Chichester, West Sussex PO19 2HF; 01243 782803

The Kennel Club, 1 Clarges Street, Piccadilly, London W1Y 8AB; 0171 493 6651

Ministry of Agriculture, Fisheries and Food (MAFF), Government Buildings (Toby Jug Site), Hook Rise South, Tolworth Tower, Surbiton, Surrey KT6 7NF; 0181 330 8289; 0645 335577 (helpline)

Monkey Sanctuary, Nr Looe, Cornwall PL13 1NZ; 01503 262532

National Association of Farriers, Blacksmiths and Agricultural Engineers, The Forge, Avenue B, Tenth Street, NAC, Stoneleigh, Warwickshire CV8 2LG; 01203 696595

National Council for Vocational Qualifications, 222 Euston Road, London NW1 2BZ; 0171 387 9898

National Dairy Council, 5–7 John Prince's Street, London W1M 0AP; 0171 499 7822

National Extension College, 18 Brooklands Avenue, Cambridge CB2 2HN; 01223 316644

National Federation of City Farms, The Green House, Hereford Street, Bedminster, Bristol BS3 4NA; 0117 923 1800

National Greyhound Racing Club Ltd, 24–28 Oval Road, London NW1 7DA; 0171 267 9256

National Horse Education and Training Company Ltd, Second Floor, The Burgess Building, The Green, Stafford ST17 4BL; 01785 227399

National Pony Society, Willingdon House, 102 High Street, Alton, Hampshire GU34 1EN; 01420 88333

National Stud, Newmarket, Suffolk CB8 0XE; 01638 663464

National Trainers Federation, 42 Portman Square, London W1H 0AP; 0171 935 2055

Northern Racing School, The Stables, Rossington Hall, Great North Road, Doncaster DN11 0HN; 01302 865462

Organic Food and Farming Centre, Education Officer, 86 Colston Street, Bristol BS1 5BB; 0117 929 9666

PDSA, Whitechapel Way, Priorslee, Telford, Shropshire TF2 9PQ; 01952 290999

Pet Care Trust, Bedford Business Centre, 170 Mile Road, Bedford MK42 9TW; 01234 273933

Pets As Therapy, National Head Office, Rocky Bank, 4–6 New Road, Ditton, Aylesford, Maidstone, Kent ME20 6AD; 01732 872222

Racing and Thoroughbred Breeding and Training Board (RTBTB), Suite 16, Unit 8, Kings Court, Willie Snaith Road, Newmarket, Suffolk CB8 9BL; 01638 560743

Recruitment and Assessment Services, Alencon Link, Basingstoke, Hampshire RG21 3JB; 01256 329222 – For applications to MAFF posts only; they do not send out careers information

Royal College of Veterinary Surgeons, Belgravia House, 62–64 Horseferry Road, London SW1P 2AF; 0171 222 2001

Royal Society for the Protection of Birds (RSPB), The Lodge, Sandy, Bedfordshire SG19 2DL; 01767 680551

RSPCA, Causeway, Horsham, West Sussex RH12 1HG; 01403 264181

Scottish Qualification Authority, Hanover House, 24 Douglas Street, Glasgow G2 1NQ; 0141 242 2214

Scottish Wildlife Trust, Cramond House, Cramond, Glebe Road, Edinburgh EH4 6NS; 0131 312 7765

Society of Master Saddlers, Kettles Farm, Mickfield, Stowmarket, Suffolk IP14 6BY; 01449 711642

Stable Lads Association, 4 Dunsmore Way, Midway, Swadlincote, Derbyshire DE11 7LA; 01283 211522

Thoroughbred Breeders' Association, Stanstead House, The Avenue, Newmarket, Suffolk CB8 9AA; 01638 661321

Universities and Colleges Admissions Service (UCAS), PO Box 40, Cheltenham, Gloucestershire GL50 3SB; 01242 227788

Universities Federation for Animal Welfare, 8 Hamilton Close, South Mimms, Potters Bar, Hertfordshire EN6 3QD; 01707 658202

The Veterinary Record, 1 Lancaster Place, Strand, London WC2E 7HR; 0171 636 6541

Whipsnade Wild Animal Park, Dunstable, Bedfordshire LU6 2LF; 01582 872171

Wildfowl and Wetlands Trust, Slimbridge Centre, Gloucestershire GL2 7BT; 01453 890333

Willing Workers on Organic Farms (WWOOF), 19 Bradford Road, Lewes, Sussex BN7 1RB; 01273 476286

Wood Green College of Animal Welfare/Animal Care Industry Lead Body, Wood Green Animal Shelters, Godmanchester, Huntingdon, Cambridgeshire PE18 8LJ; 01480 831177

World Wide Fund for Nature (WWF), Panda House, Weyside Park, Godalming, Surrey GU7 1XR; 01483 412303

Northern Ireland

Conservation Volunteers Northern Ireland, 159 Raven Hill Road, Belfast, BT6 0BP; 01232 645169

Department of Agriculture, Dundonald House, Upper Newtownards Road, Belfast BT4 3SB

The Equestrian Centre, The Drumlin, 49 Ballyworfy Road, Hillsborough, County Down BT26 6LR; 01846 682539

Guide Dogs for the Blind, Lanesborough House, 15 Sandown Park South, Belfast BT5 6HE; 01232 471453

Republic of Ireland

Blue Cross Dublin Clinic, 9 Dartmouth Terrace, Ranelagh, Dublin 6; 01149 71985

Conservation Volunteers Ireland, Royal Dublin Society, Ballsbridge, Dublin 4

Irish Guide Dogs for the Blind, Mobility Training Centre, Model Farm Rd, Cork; 01218 70929

Irish National Stud Company Ltd, Tully, Kildare, County Kildare; 01045 521251

Irish RSPCA, 300 Lower Rathmines Road, Dublin 6; 01149 77222

Racing Apprentice Centre of Education, Curragh House, Dublin Road, Kildare; 01455 21305/22468

The Veterinary Council of Ireland, 53 Lansdowne Road, Ballsbridge, Dublin 4; 01166 84402

Willing Workers on Organic Farms (WWOOF), c/o Annie Sampson, Kilcornan, Kilkishen, Clare

15 Further reading

Vets
The Veterinary Profession (free), *Careers for Veterinary Surgeons* (£4.00), *Do You Want to Be a Vet?* (£16.00 Video); all from The Royal College of Veterinary Surgeons
The Veterinary Science Degree Course Guide (£4.99) CRAC Publications

Veterinary nursing
A recommended reading list and a number of free careers leaflets are available from the British Veterinary Nursing Association on receipt of an sae.

Animal welfare and conservation charities
Animal Rescue Directory (£2.95) Hand to Paw
Who's Who in the Environment England (£12.50), Wales (free with sae + 64p p+p), Scotland (£6.00) The Environment Council
Directory of Environmental Courses (available in careers libraries) The Environment Council
Careers in Animal Welfare, A Career as an Inspector, A Career as a Kennel Assistant (free on receipt of a large sae envelope with 50p in stamps) RSPCA

Working with horses
A Career in the Horse Industry (£9.95) Kenilworth Press, Addington, Buckingham MK18 2JR
British Equestrian Directory (£14.95) EMC & BETA; 0113 289 2267
The Sporting Life, Racing Post and *Horse and Hound* – journals available from most newsagents
Where to Ride (£6.95) British Horse Society
Careers in Horse Racing (Price unavailable at time of press) Racing & Thoroughbred Breeding Training Board

Further reading

Working with dogs
Obedience Competitor supply a recommended reading list for potential dog trainers. Their address is Long Meadow, Mooredges, Thorne, Doncaster DN8 5RY. Alternatively, your local library may be able to order one of the following; *Training the Family Dog* by John Holmes; *Living with Dogs* by George Summers; *Understanding Your Dog* by Peter Griffiths and *Heel Away Your Dog* by Charles Wilde.
How To Work with Dogs (£9.99), The Pet Behaviour Centre, Upper Street, Defford, Worcestershire WR8 9AB.

Pets
The Pet Care Trust will send out free leaflets on working in a pet shop and dog grooming.

Zoos
The Federation of Zoological Gardens of Great Britain and Ireland will send you a number of leaflets regarding conservation, education, welfare and careers.

Agriculture
Careers in Landbased Industries (free) ATB Landbase/Warwickshire Careers Service or from careers offices.
Directory of Courses in Land Based Industries (£19.95), Farming Press Books, 01473 241122

Index

Accelerated Modern Apprenticeships 74
Agricultural Training Board (ATB) 61
Agricultural Wages Board 63
agriculture 58–65, 69–70
  Modern Apprenticeships 63, 74
  pay 63
  specialisation 61
  training 61–3, 73
  work experience 61
  see also city farms; fish farming; organic farming
Animal Care College 38, 82
animal conservation 22–3, 68
animal grooming 47–9
animal health officers 8
animal photography 49–50
Animal Rescue Directory 1996/7 17, 23
animal sanctuaries 23
animal welfare 9
  charities 17–24
  organisations 68
apprenticeships 48
armed services 33–5
Association of British Riding Schools (ABRS) 30–32
Association of British Wild Animal Keepers (ABWAK) 57
ATB Landbase 61, 62, 64, 74

Birds 23
The Blue Cross 21–2
British Dog Groomers Association 48
British Horse Society (BHS) 29, 30
British Institute of Professional Dog Trainers 40
British Racing School 25, 26
British Veterinary Nursing Association (BVNA) 12, 13
BTEC Certificate 73
BTEC Diploma 73
BTEC Higher National Diploma 73

Cage and Aviary Birds 57
Careers for Veterinary Surgeons 5
city farms 64–5
college courses 76–81
correspondence courses 82
credit schemes 71

dairy farming 59–60
dairy herdsman 59–60
Directory of Courses in Land Based Industries 62
Directory of Environmental Courses 1997–1999 22
Directory of Turf 27
Do you want to be a vet? 5
dog beauticians 48
dog grooming, case study 48–9
dog training 39–41
dogs, working with 36–45, 69
  see also guard dogs; guide dogs; hearing dogs for the deaf; kennel work; police dogs

Environment Council 22

farmers and farming, *see* agriculture;

Index

city farms; fish farming; organic farming
Farmers Guardian 61
Farmers Weekly 61
farriers 32–3
Farriers' Registration Council 32, 33
Farriery Apprenticeship Scheme 33
Farriery Training Service (FTS) 33
Federation of Zoological Gardens of Great Britain and Ireland 51, 52, 57
fish farming 64

General National Vocational Qualifications (GNVQs) 72–3
greyhound racing 39
guard dogs 45
guide dogs 42–4
    kennel staff 42–3
    mobility instructor 43–4
    trainers 43
Guide Dogs for the Blind Association 42

Hearing Dogs for the Deaf 44–5
Horse and Hound 25
Horse Protection Scheme 21
horses, working with 25–35, 69
Modern Apprenticeships 73
Horses in Training 25
Household Cavalry 34

Institute of Fisheries Management 64
International Zoo News 57

jockeys 26

kennel work 36–8
    standards 37–8
    training 38
King's Troop Royal Horse Artillery 33–4

Local Enterprise Company (LEC) 74

Meat Hygiene Regulations 9
Ministry of Agriculture, Fisheries and Food (MAFF) 8, 10
mobility instructor 43–4
Modern Apprenticeships 69, 73
    Accelerated 74
    agriculture 63, 74
    horse industry 31, 73
The Monkey Sanctuary 24
mounted police 34–5

National Association of Grooms (NAG) 27
National Certificate (NC) 73
National Dairy Council 60
National Diploma (ND) 73
National Extension College 82
National Federation of City Farms 65
National Greyhound Racing Club 39
National Horse Education and Training Company 31, 73
National Pony Society 28
National Small Animal Care Certificate 38
National Stud 27, 28
National Trainers Federation 25
National Vocational Qualifications/Scottish Vocational Qualifications (NVQs/SVQs) 26, 28, 31, 36, 38, 63, 69, 71–3
Northern Racing School 25

organic farming 64, 70

People's Dispensary for Sick Animals (PDSA) 9, 20–21
Pet Care Trust 69, 82
pet shops 46–7
pet trade 69

Index

photography, *see* animal photography
police dog handlers   41–2
   training   41–2
police service   34
pre-vocational courses   71

qualifications   71–4

Racing and Thoroughbred Breeding Training Board (RTBTB)   26
Racing Post   25
RATEL   57
riding schools   29–30
Royal Army Veterinary Corps (RAVC)   34
Royal College of Veterinary Surgeons (RCVS)   5, 8, 13
Royal Society for the Prevention of Cruelty to Animals (RSPCA)   18–20
Royal Society for the Protection of Birds (RSPB)   23

Scottish Certificate and Diploma courses   73
Scottish Vocational Qualifications, *see* National Vocational Qualifications/Scottish Vocational Qualifications (NVQs/SVQs)
shepherd   59
show jumping   28
Soil Association   64
Sporting Life   25
stable lads   25–7
State Veterinary Service (SVS)   8–9
stud work   27–8

Talland School of Equitation   29

TGWU   27
Thoroughbred Breeders Association   28
Training and Enterprise Council (TEC)   74
training establishments   75–82

Veterinary Field Service (VFS)   8–9
veterinary nurses (VNs)   12–16, 71
   prospects   15
   qualifications   12–13
   training   13–14
veterinary officers   8
Veterinary Profession   5
veterinary science   5
   degree courses   5, 75–6
Veterinary Science Degree Course Guide, The   5
veterinary surgeons   5–11, 34, 68
   inspection duties   9
   prospects   8–10
   qualities   11
   salaries   10–11
vocational training   71

Willing Workers on Organic Farms (WWOOF)   61

zoo keeping   52
   further qualifications   56
   qualifications and training   52–3
   recruitment   57
   routine work   54–5
   salaries   56
   typical day   55
   visitor activity   55–6
   working conditions   53–5
Zoological Society of London   51
zoos   51–7